AF324609

WATER POLLUTION

Edited by **Nuray Balkis**

INTECHOPEN.COM

Water Pollution

Edited by Nuray Balkis

CBS, Edition **2016**

Published by InTech

Janeza Trdine 9, 51000 Rijeka, Croatia

Publishing Process Manager Anja Filipovic

Technical Editor Teodora Smiljanic

Cover Designer InTech Design Team

Additional hard copies can be obtained from orders@intechopen.com

Water Pollution, Edited by Nuray Balkis
p. cm.
ISBN 978-953-307-962-2

Contents

Preface

Water pollution is a major global problem that requires ongoing evaluation and revision of water resource policy at all levels (from international down to individual aquifers and wells). It has been suggested that it is the leading worldwide cause of deaths and diseases, and that it accounts for the deaths of more than 14,000 people daily. In addition to the acute problems of water pollution in developing countries, industrialized countries continue to struggle with pollution problems as well. Water is typically referred to as polluted when it is impaired by anthropogenic contaminants and either does not support a human use, such as drinking water, and/or undergoes a marked shift in its ability to support its constituent biotic communities, such as fish. Natural phenomena such as volcanoes, algae blooms, storms, and earthquakes also cause major changes in water quality and the ecological status of water. Most water pollutants are eventually carried by rivers into the oceans. In some areas of the world, the influence can be traced up to one hundred miles from the mouth by studies using hydrology transport models.

The specific contaminants leading to pollution in water include a wide spectrum of chemicals, pathogens, and physical or sensory changes, such as elevated temperature and discoloration. While many of the chemicals and substances that are regulated may be naturally occurring (calcium, sodium, iron, manganese, etc.), the concentration is often the key in determining what is a natural component of water, and what is a contaminant. High concentrations of naturally-occurring substances can have negative impacts on aquatic flora and fauna. Oxygen-depleting substances may be natural materials, such as plant matter (e.g. leaves and grass) as well as man-made chemicals. Other natural and anthropogenic substances may cause turbidity (cloudiness) which blocks light and disrupts plant growth, clogging the gills of some fish species. Many of the chemical substances are toxic. Pathogens can produce waterborne diseases in either human or animal hosts. Alteration of water's physical chemistry includes acidity (change in pH), electrical conductivity, temperature, and eutrophication. Eutrophication is an increase in the concentration of chemical nutrients in an ecosystem to an extent that increases in the primary productivity of the ecosystem. Depending on the degree of eutrophication, subsequent negative environmental effects such as anoxia (oxygen depletion) and severe reductions in water quality may occur, affecting fish and other animal populations.

This book includes scientific papers about water pollution, control and solution. Thus, it forms a significant base for future studies.

Dr. Nuray Balkis
Istanbul University
Institute of Marine Science and Management
Department of Chemical Oceanography
Istanbul, Turkey

Periphyton and Earthworms as Biological Indicators of Metal Pollution in Streams of Blantyre City, Malawi

C. C. Kaonga and M. Monjerezi
University of Malawi
Malawi

1. Introduction

Pollution of water and soils arises from overburdens of mines, application of fertilizers and pesticides, industrial effluents and sewage sludge (Alloway & Ayres, 1997), among others. In heavily contaminated soils and water, there is a decrease in the population, growth and function of biota. In most cases all biological indices of environmental health (fish, invertebrates and algae) decline as pollution intensity increases (Cuffney et al., 2000; Hill et al., 2000; Khan, 1990). The identification of plant and animal species with the ability to accumulate metals is therefore of interest for the purposes of environmental monitoring (Chukwuma, 1998; Manly, 1996).

Earthworms and periphyton (attached algae) have been utilized as indicators of pollution of soils and water with metals (Ireland, 1983; Holan et al., 1993; McCormick & Cairns, 1994; Ramelow, 1987; Jin-fen et al., 2000). Despite being very small, experiments unequivocally demonstrate that algae sequester heavy metals by complexation to phytochelatins (Gekeler et al., 1988), which is an identical mechanism as higher plants. In the context of biomonitoring, earthworms act as quantitative monitors of total-soil metal and also estimators of ecologically significant soil metal concentration (Morgan & Morgan, 1988). Earthworms are important components of the soil system mainly because of their favourable effects on soil structure and function which include increasing soil fertility by formation of an organic matter layer in topsoil. The most widely studied earthworm species are *Eisenia fetida*, *Eisenia Andrei*, *Lumbricus terrestris* and *Lumbricus rubellus* (Georgescu & Weber, 2007).

In Malawi, several studies have confirmed the presence of heavy metals in water and soils. Kadewa et al. (2001) found levels of copper, cadmium and chromium in soils fertilized by sewage sludge from Soche waste water treatment plant in Blantyre to be higher than the range for critical concentration for sludge amended soils. Sajidu et al. (2007) found that the levels of lead, cadmium, iron, manganese, zinc, chromium and nickel in streams in the city of Blantyre were much higher than World Health Organisation (WHO) safe limits for drinking water in all sampled streams after they had passed through industrial areas. Lakudzala et al. (1999) found that at some points on Mudi, Likhubula and Shire Rivers, the iron and lead levels exceeded WHO guideline limits. However, not much has been done on the use of biological indicators to assess the state of the environment. In addition, studies that assess levels of heavy metals in biota are lacking because they only target either invertebrates or aquatic plants only. Compounds of heavy metals in earthworms may be transferred to other species at higher trophic levels and may be lethal to earthworm consumers (Hui, 2002; Ireland and Richards, 1977; Ma, 1982; Vyas et al., 2000). Monitoring programs, with a well-founded scientific base and defined management outcomes, using biological indicators (such as algae, fungi, earthworms and other microorganisms), will expand our knowledge of river/aquatic function (Burns & Ryder, 2001; Khosmanesh et al., 1996). This work reports on the levels of potentially harmful elements in the streams, wastewater and stream bank soils. In addition, aspects of metal accumulation in earthworms and green algae and their potential for biomontitoring are presented.

2. Materials and methods

2.1 Study area

Malawi is situated in South East Africa (Fig. 1a). This study was conducted in the City of Blantyre (Fig. 1b), the commercial and industrial city of Malawi. The City has eight designated industrial areas namely *Makata*, Ginnery corner, *Maselema, Limbe, Chirimba*, south *Lunzu, Maone* and *Chitawira*, with south *Lunzu* still under development (Fig. 1c). All the industrial sites are located along the banks of the main streams in the City. *Makata* industrial site lies between *Mudi* and *Nasolo* streams, Ginnery corner is along *Mudi* stream, *Maselema* is along *Naperi* River and *Chirimba* is along *Chirimba* stream (Fig. 1c). The sampling points fell into two major categories, which were; streams and wastewater treatment plants (WWTP). The waste water treatment plants were included because their effluent is released into the streams. Most of these streams pass through the major industrial areas except for Namangunda, which passes by a dumpsite and Michiru, which originates from a forest reserve and does not pass through the industrial sites (Fig. 1c).

2.2 Sample collection

All samples were collected in wet (November - February) and dry (July - October) seasons, to capture the effects of seasonal variation, from the selected streams and WWTPs in Blantyre City (Fig. 1). A total of eighteen periphyton (algae) samples were collected for each season. The samples were collected in 100 mL plastic bottles (Diatoms for Assessing River Ecological Status (DARES), 2004). The algae samples were chilled in a refrigerator pending analysis (New South Wales (NSW), 2002). The periphyton was identified as *Spirogyra aequinoctialis*.

Water samples were collected at an area where samples of *S. aequinoctialis* were collected. A total of forty three (43) water samples were collected for each season. Grab sampling method was used in the collection of water samples both upstream and downstream of a designated industrial area. At each sampling point, water samples were collected in triplicates for heavy metal analysis and a single sample for pH analysis. The samples were collected and stored in 1 L pre-cleaned new polyethylene bottles. Water samples for determination of metal were acidified to pH < 2 by adding concentrated nitric acid (Analytical Reagent (AR)) (American Public Health Association (APHA), 2005).

A total of forty-six (46) earthworm and soil samples were collected in both seasons. The earthworms were collected in 400 mL plastic bottles (at least three individuals per sampling site) into which a few holes were poked on the lid (Ecological Monitoring and Assessment Network (EMAN), 2004). Earthworm casts were used to find possible earthworm locations. Soil (stream sediment) samples were collected where earthworms were found using a soil auger. Soil samples were collected within the top soil range (0-20 cm) since most of the earthworms were found in this region. Soil horizons could not be distinguished in all these sampling points. Five augerings were collected at each site and were mixed in a bucket before sub sampling (quartering) (Anderson and Ingram, 1993). The samples were collected in plastic bags.

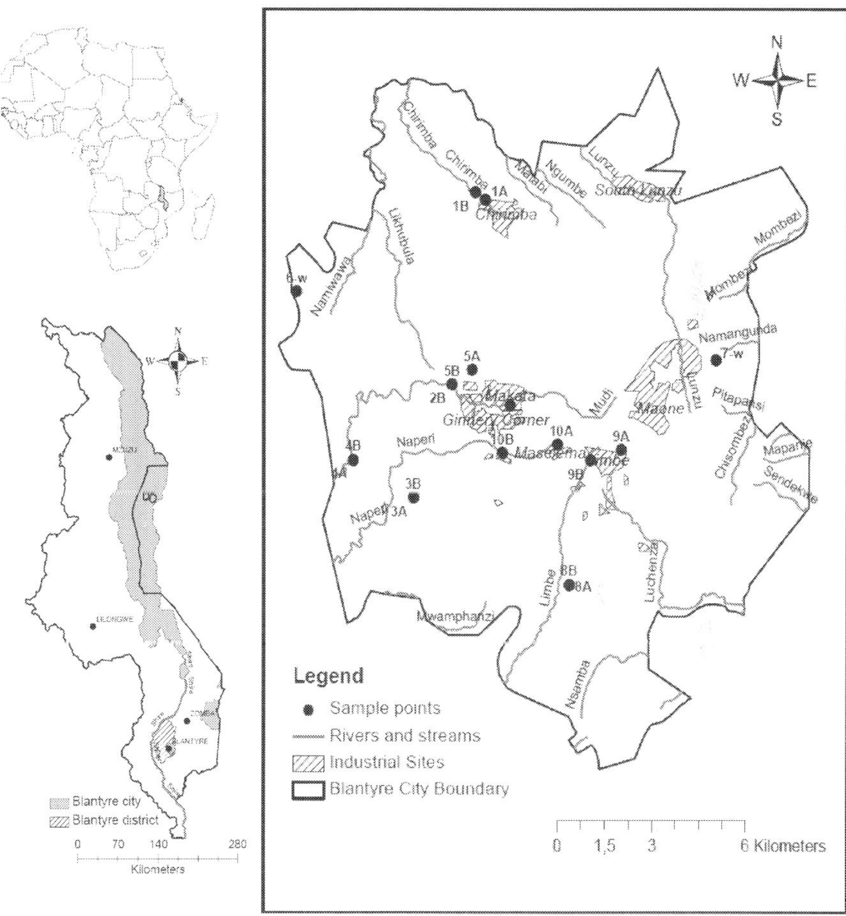

Fig. 1. Maps showing the location of Malawi in Africa, Blantyre city in Malawi and sampling points in the City of Blantyre. Sample IDs are explained in Table 1.

2.3 Analysis of water and wastewater samples

pH was measured immediately after sampling using Orion Research digital ionalyzer 601A and Metrohm 744 pH meters (ISO 10523-1:1994). Water samples were digested using concentrated nitric acid (AR). 50 mL of the sample was transferred to a beaker to which 5 mL concentrated nitric acid was added and brought to a boil on a hot plate to the lowest volume possible (15 to 20 mL). Filtration was done after digestion. The filtrate was then diluted to volume with distilled water in a 50 mL volumetric flask (APHA, 2005). Total concentrations of Mn, Cd, Cr, Cu, Pb, Ni, Fe, Zn were determined using flame atomic absorption spectroscopy (Perkin Elmer, Analyst 700; APHA, 2005).

2.4 Analysis of soil samples

The soil samples were air dried, ground in a mortar and passed through a 2 mm sieve. 5.0 g of the dry sieved soil sample was heated with 10 mL concentrated nitric acid (AR) for 45 minutes. The soil samples were then dried, re-dissolved in 5 mL aqua regia (3:1 conc. HCl (AR) and conc. nitric acid (AR)) and filtered. Total concentrations of Mn, Cd, Cr, Cu, Pb, Ni, Fe, Zn were determined using flame atomic absorption spectroscopy (Perkin Elmer, Analyst 700 and Buck Scientific AAS model 200A; Bamgbose et al., 2000). The total metal concentrations were expressed as mg/kg dry weight of soil (mg/kg dw).

Soil organic matter was determined using the Walkley – Black method (Walkley and Black, 1934). Briefly, the soil samples were ground using a mortar and then passed through a 0.5 mm sieve. 1.00 g soil was mixed with 10 mL of 1N potassium dichromate (AR) solution and 15 mL concentrated sulphuric acid (AR), whilst shaking. The mixture was then shaken for a further one minute and left to stand for thirty minutes. Then 150 mL distilled water and 5 mL concentrated phosphoric acid (AR) were added whilst shaking. After cooling, the mixture was titrated against 0.5 N ferrous ammonium sulphate (AR) solution, with 1 mL diphenylamine indicator. The colour change was from deep blue to dark green. Similarly triplicates of blank titrations were carried out. Where the volume of 0.5 N ferrous ammonium sulphate (AR) solutions was less than 3 mL, the determinations were repeated using 0.5 g soil. The percentage organic matter was calculated using the following equation:

$$\%OM = 1.729 \times 0.0031 \times 100 \times F \times \left[\frac{Me_{K_2Cr_2O_7} - Me_{Fe(NH_4)_2.6H_2O}}{mass\ (g)\ of\ air\ dried\ soil} \right] \qquad (1)$$

where, F = Correction factor (1.33) and Me = Normality of solution $\times$ volume (mL) of solution used.

Soil pH was determined using glass electrode pH meters (model 601A Orion Research digital ionalyzer and model 744 Metrohm) in a 1:5 (V/V) of soil in water (pH-H_2O).

2.5 Analysis of metals in earthworm samples

The earthworms were identified as *Aporrectodea icteria* after being rinsed with distilled water. Only reproductively mature earthworms were identified because of presence of a clitellum. Then they were placed on moist filter papers and put in glass Petri dishes (one individual per dish) and kept at 10º C for 24 hrs in order to purge the soil in the gut. They were then rinsed slightly with distilled water and then stored frozen and then freeze-dried. Gut contents remaining in some earthworms were removed manually. 3.0g of thawed and dried earthworm sample was heated with 2 mL concentrated HNO_3 (AR), filtered and made up to 50 mL with distilled water. The metal contents were determined by running samples on AAS as for the soil samples. The metal concentrations were expressed as mg/kg dry weight of earthworm (mg/kg dw).

2.6 Analysis of metals in *S. aequinoctialis* samples

Periphyton samples were air dried (Hoffman, 1996). The air dried samples were then dry-ashed in a furnace after adding nitric (AR) and hydrochloric acid (AR) (Association of Official Analytical Chemists (AOAC), 1990). Thereafter the sample was made up to 50 mL with distilled water in a volumetric flask. The samples were prepared in triplicates and blank and standard samples were used to check accuracy of analysis. The concentration of heavy metals was determined by running samples on AAS (Perkin Elmer, Analyst 700).

2.7 Quality control

To ensure quality control of sampling and analysis, a number of procedures were followed. Firstly the sampling devices were carefully chosen so that they should not contaminate the samples. If the same apparatus were to be used for the next sampling process, they were thoroughly cleaned and rinsed with distilled water. For water samples in which heavy metals were to be determined, acidification was done to pH less than 2 to avoid adsorption of metals on the sides of the sampling containers. In the measurement of volume a pipette which is more accurate was used rather than measuring cylinders. Soil samples were prepared away from the rest of the samples since dust could easily contaminate the other samples. All samples for heavy metal analysis were determined in triplicates. Analytical reagents were used for all procedures rather than general purpose reagents. In addition, the following minimum laboratory quality control measures (United States Environmental Protection Agency (US-EPA), 2010) for the instruments used in the analysis and also samples were followed;
a.   Initial calibration; This was done prior to analysis of samples (minimum three concentration levels for every compound and an instrument blank).
b.   Continuing calibration; This was done once per 10 samples (mid-level standard containing all compounds) and a continuing calibration blank.
c.   Method blank; This was done once per digestion or extraction set.
d.   Soil and water samples were preserved at 4°C and analysed within 28 days.

3. Results

3.1 Metal concentrations in stream water, wastewater and *S. aequinoctialis*

Table 1 provides the levels of determined metals in stream water and wastewater samples. The corresponding World Health Organization (WHO) drinking water guidelines (WHO, 2006) and the Malawi Standard (MS 214) (MBS, 2005) for the parameters are also included in Table 1. The levels of metals determined in *S. aequinoctialis* samples are provided in Table 2. In *S. aequinoctialis* samples, concentrations of manganese, cadmium and copper were significantly higher ($p < 0.05$) in the dry season than in the rainy season. There were, however, no significant seasonal differences in the levels of lead, zinc, chromium and nickel.

Chromium and copper were not detected in all samples in the rainy season, but they were measured in levels of up to 0.419 mg/L and 0.076 mg/L, respectively, in the dry season (Table 1). For both seasons, the determined levels of zinc and copper were below MS 214 and WHO water quality guidelines, whereas levels of nickel and cadmium were above these guidelines (Table 1). 17% of the samples had chromium levels above the MS 214 (0.05-0.1 mg/L) and WHO (0.05 mg/L) water quality standards. For lead, 44 % and 61 % of the sampled points contained lead levels above MS 214 (0.01 – 0.05 mg/L) and WHO (0.01 mg/L) drinking water standards in the rainy and dry seasons, respectively. In the case of manganese, water quality standards were only exceeded at Mangunda stream, in the rainy season. In the dry season, however, 83% and 17% of the sampling points showed manganese levels above the MS 214 and WHO drinking water quality guidelines, respectively (Table 1).

3.2 Metal content in soils and *A. icteria*

Table 3 and Table 4 provide results of soil and earthworm sample analyses, respectively. The metal content in the assessed soil sites is low in comparison to guideline values in several European countries. Levels of Cd (rainy season), Pb, Cr, Cu, Zn and Ni (both seasons) were below their respective England toxic limits (0.06 mg/kg for Cd, 10 mg/kg for Pb, 50 mg/kg for Zn , 20 mg/kg for Cr and 40 mg/kg for Ni; Bohn et al, 1985), Swiss guide levels (0.8 mg/kg dry soil for Cd, 50 mg/kg dry soil for Pb; OIS, 1998) and the Netherlands target levels (85 mg/kg for Pb, 36 mg/kg for Cu, 140 mg/kg for Zn, 100 mg/kg for Cr, 35 mg/kg for Ni; Alloway and Ayres, 1997). However, for the dry season, 33% of the soil samples were above the England toxic level (0.06 mg/kg; Bohn et al., 1985).

The internal concentrations of Cd, Cu, Pb, Zn and Cr were below the levels that show significant changes in (sub-) lethal endpoints for earthworms (see e.g. Langdon et al, 2001; Spurgeon and Hopkin, 1999; Spurgeon et al, 2000). There were significantly higher concentrations of Cd in *A. icteria* than in the soils, but significantly lower values of Mn, Fe, Pb, Cr, Zn and Cu in the earthworm than soils ($p < 0.05$). There was no significant difference in the concentrations of Ni in soils and earthworms ($p > 0.05$). The effect of seasonality varies among the studied metals. In the soils, levels of Mn were significantly higher in dry season than the rainy season ($p < 0.05$), but there were no significant differences between the seasons for the values of total soil concentrations of Cd, Cu, Zn, Pb, Cr and Ni ($p > 0.05$). pH was significantly higher in the rainy season than the dry season ($p < 0.05$), but there were no significant differences in soil OM content between the seasons ($p > 0.05$). In *A. icteria*, levels of Cd and Cr were significantly higher in dry season than the rainy season ($p < 0.05$), but there were no significant differences between the seasons for the values of concentrations of Mn, Cu, Zn, Pb, Ni and Ca ($p > 0.05$).

4. Discussion

4.1 Potential sources of metal pollution

Pearson correlations were calculated to find empirical inter-relationships between the chemical parameters. Correlation between chemical parameters may indicate similar origins or conceptual relationships, as well as common governing factors. In the soil samples, concentrations of Cr were significantly correlated with Zn, Cu and Pb in the rainy season and with Pb in the dry season (Table 5). The strong association of these metals with each other indicates their anthropogenic origin. Organic matter content strongly affects the soil content of Cd, Zn (rainy season) and Cr (Table 5).

The presence of heavy metal pollution in the streams of Blanytre City has been reported upon by Sajidu et al (2007) and Kuyeli (2007) and both studies pointed at industrial activities as the possible sources of pollution. Kuyeli (2007) reported Cd in effluent from printing (0.034 mg/l), textiles (0.034 mg/l), motor oil (0.025 mg/l), battery (0.019 mg/l) and abattoir industry (0.06 mg/l) in the dry season; Cr in effluent from match stick production (41.59 mg/l in the dry season and 56.12 mg/l in the rainy season); Cu in the range 0.026 mg/l (battery manufacturer) to 2.00 mg/l (Paint industry); Zn in effluent from battery manufacturer (30.83 mg/l) and match stick production (15.51 mg/l) in the rainy season and 18.97 mg/l (match stick), 13.9 mg/l (battery) and 14.4 mg/l (fertiliser manufacturer) in the dry season; Pb in paint (1.29 mg/l), printing (2.60 mg/l) in the dry season and match stick (0.465 mg/l) and printing (0.233 mg/l) in the rainy season. Sajidu et al (2007) reported a significant increase in the levels of Pb, Cd, Cr, Fe, Cu, Ni and Mn in the same Blantyre streams after passing through an industrial site. The results from this study show enrichment of the heavy metals (Zn, Cd, Cr and Pb) in most streams over that of Michiru stream, which is in a forest reserve. In Malawi, cadmium is present in coatings on steel and also in batteries and potassium dichromate ($K_2Cr_2O_7$) is used as a raw material for producing match-heads. Copper compounds are used in textile, print and paint industry for pigmentation whereas Pb is used as a pigment, dispersing and drying agent in the print and paint industry. In match stick production lead oxide is used to give the scarlet colour of the match.

4.2 Accumulation of metals in *S. aequinoctialis*

Calculated bioconcentration factors (BCF) show that *S. aequinoctialis* accumulated heavy metals in the order Mn> Zn>Cu> Pb (Table 6). *S. aequinoctialis* had significantly higher ($p <$ 0.05) levels of lead, copper, zinc and manganese than the corresponding water samples, in both seasons. There were no significant differences in levels of chromium between the algae samples and water samples whereas the differences were season dependent for the other metals. Water samples had high cadmium levels in the rainy season while in the dry season the levels were higher in *S. acquinoctialis*. For nickel, water samples indicated significantly higher nickel levels than *S. aequinoctialis* ($p < 0.05$), in the rainy season, but there was no significant difference in the dry season ($p > 0.05$).

There were strong correlations between water and algae metal contents for Cu ($r = 0.73$; $p <$ 0.05; Fig. 2a) and Cr ($r = 0.65$; $p < 0.05$) in the rainy season. A low correlation for Mn ($r = 0.40$; $p < 0.05$) was also obtained for the dry season. There were no correlations for the other metals. There is an established consensus in the literature that brown and green algae are capable of biosorption of metals from their environment (Davis et al., 2003; Rajfur et al., 2010). They have thus been used in biomonitoring of heavy metals mostly in marine environments (Filho et al., 1999; Żbikowski et al., 2007; Akcali and Kucuksezgin, 2011). This study is in agreement with these studies and adds to the knowledge of heavy metal accumulation of green algae in a fresh water environment. Heavy metal levels in algae species are dependent both on environmental parameters (salinity, temperature, pH, light, oxygen, nutrient concentrations, complexing agents) and on the structural differences among the algae species (Garnharm et al., 1992; Favero et al., 1996).

4.3 Metal accumulation in *A.icteria*

Concentrations of Cu, Zn (rainy season) and Cd (dry season) in the soil were significantly correlated with the concentrations in *A.icteria* (Fig. 2b-d); Table 5). These metals also show correlations of varying strength with soil OM content (Table 5). The calculated bioconcentration factors (BCF) show that *A. icteria* accumulated heavy metals in the order Cd > Zn = Ni > Pb > Cu = Cr (Table 7), consistent with data from other similar studies (e.g.

ID	Sampling Area	Wet Season								Dry Season							
		pH	Mn	Cd	Cu	Zn	Pb	Cr	Ni	pH	Cd	Mn	Cu	Zn	Pb	Cr	Ni
1A	Chirimba at Cori	7.53	[1]ND	0.073 ±0.005	ND	0.502 ±0.056	0.037 ± .001	ND	0.398 ±0.013	6.81	0.037 ±0.005	0.035 ±0.014	0.046 ±0.037	0.295 ±0.240	0.026 ±0.015	ND	0.420 ±0.009
1B	Chirimba at Machinjiri	7.22	ND	0.073 ±0.004	ND	0.558 ±0.164	0.035 ±0.002	ND	0.391 ±0.004	7.05	0.041 ±0.003	0.155 ± 0.07	0.023 ±0.002	0.148 ±0.047	0.108 ±0.004	ND	0.405 ±0.004
2A	Mudi at MDI	7.54	0.060± 0.028	0.111 ±0.031	ND	1.494 ±0.002	0.038 ±0.014	ND	0.329 ± 0.04	7.58	0.052 ±0.137	0.244 ± 0.01	0.006 ±0.022	0.116 ±0.057	0.079 ±0.014	ND	0.349 ±0.103
2B	Mudi at SRN	7.39	ND	0.085 ±0.003	ND	2.614 ±.521	0.064 ±0.048	ND	0.347 ±0.006	7.19	0.047 ±0.011	0.178 ± 0.04	0.045 ±0.109	0.102 ±0.019	0.091 ±0.043	0.395 ±0.085	0.573 ±0.034
3A	Soche WWTP raw	7.62	ND	0.089 ±0.003	ND	0.703 ±0.183	0.042 ±0.016	ND	0.387 ±0.016	7.21	0.111 ±0.013	0.365 ±0.027	0.065 ±0.892	0.233 ±0.031	0.110 ±0.007	ND	0.234 ±0.008
3B	Soche WWTP effluent	7.79	ND	0.087 ±0.003	ND	0.711 ±0.187	0.058 ±0.013	ND	0.392 ±0.007	7.25	0.087 ±0.009	0.384 ±0.029	0.054 ±0.003	0.195 ±0.157	0.014 ±0.010	ND	0.101 ±0.003
4A	Blantyre WWTP raw	7.2	ND	0.081 ±0.002	ND	0.674 ±0.034	0.047 ±0.011	ND	0.426 ±0.029	6.98	0.092 ±1.002	0.435 ±0.011	0.076 ±0.153	0.173 ±0.038	0.061 ±0.006	0.297 ±0.058	0.505 ±0.007
4B	Blantyre WWTP effluent	7.55	ND	0.082 ±0.003	ND	0.742 ±0.111	0.034 ±0.008	ND	0.409 ±0.030	7.04	0.075 ±0.018	0.453 ±1.034	0.044 ±0.113	0.135 ±0.061	0.052 ±0.032	0.014 ±0.013	0.317 ±0.003
5A	Nasolo at BNC	7.53	ND	0.082 ±0.004	ND	1.079 ±0.134	0.069 ±0.039	ND	0.497 ±0.007	7.13	0.079 ±0.017	0.42 ±0.029	0.038 ±0.293	0.133 ±0.012	0.092 ±0.041	0.025 ±0.004	0.365 ±0.008
5B	Nasolo at SRN	8.8	ND	0.08 ± 0.008	ND	0.951 ±0.133	0.074 ±0.015	ND	0.451 ±0.091	5.98	0.098 ±1.016	0.457 ±0.018	0.064 ±1.113	0.159 ±0.050	0.048 ±0.011	0.036 ±0.011	0.515 ±0.110
6	Michiru	5.99	ND	0.086 ±0.002	ND	0.526 ±0.038	0.012 ±0.003	ND	0.413 ±0.024	7.37	0.014 ±0.001	0.056 ±0.001	ND	0.139 ±0.032	ND	ND	0.113 ±0.001
7	Mangunda	7.79	0.530 ±0.121	0.09 ± 0.003	ND	0.503 ±0.066	0.098 ±0.014	ND	0.394 ± 0.02	7.32	0.102 ±0.007	0.489 ±0.006	0.013 ±1.018	0.151 ±0.069	0.102 ±0.017	0.419 ±0.003	0.578 ±0.012
8B	Limbe WWTP effluent	10.13	ND	0.077 ± .002	ND	0.675 ± 0.09	0.074 ±0.002	ND	0.349 ±0.032	9.68	0.065 ±0.104	0.511 ±0.015	ND	0.291 ±0.285	ND	ND	0.236 ±0.065
8A	Limbe WWTP raw	7.4	ND	0.08 ± 0.004	ND	0.629 ±0.056	0.065 ±0.013	ND	0.215 ±0.035	6.84	0.018 ±2.101	0.626 ±0.041	0.013 ±1.018	0.264 ±0.113	0.04 ±0.001	ND	0.305 ±0.008
9A	Limbe at Mpingwe	7.19	ND	0.07 ± 0.002	ND	0.562 ±0.019	0.033 ±0.019	ND	0.433 ±0.044	7.18	0.049 ±0.131	0.0614 ±1.114	ND	0.403 ±0.332	0.23 ± 0.019	ND	0.155 ±0.019
9B	Limbe at Highway	7.46	ND	0.072 ±0.004	ND	0.633 ±0.116	0.089 ±0.006	ND	0.416 ±0.012	7.12	0.168 ±0.008	0.095 ±0.180	0.016 ±0.007	0.172 ±0.028	0.083 ±0.015	0.037 ±0.016	0.434 ±0.264
10A	Naperi at Rainbow	7.06	ND	0.081 ±0.003	ND	0.714 ±0.103	0.011 ±0.002	ND	0.432 ±0.009	7.28	0.585 ±0.012	0.089 ±1.982	ND	0.143 ±0.035	0.039 ±0.012	ND	0.475 ±0.106
10B	Naperi at Moi	7.01	ND	0.082 ±0.001	ND	0.621 ±0.064	0.038 ±0.004	ND	0.318 ±0.004	7.19	0.464 ±0.004	0.092 ±1.089	0.025 ±1.015	0.119 ±0.002	0.057 ±0.011	ND	0.405 ±0.011
	[2]MS 214		0.05-0.1	0.003 – 0.005	0.5 – 1	3.0– 5.0	0.01– 0.05	0.05– 0.1	0.05 –0.15								
	[3]WHO		0.5	0.003	2	3	0.01	0.05	0.02								

[1]Not detected
[2] Malawi Bureau of standards (Standards for drinking water)
[3] World Health Organisation (Standards for drinking water)
Values are in the form of mean ± standard deviation

Table 1. Levels of Ni, Cu, Fe, Pb, Cr, Cd, Mn and Zn in water samples (concentrations are in mg/L)

Sampling point	Rainy Season							Dry season						
	Mn	Cd	Cu	Zn	Pb	Cr	Ni	Mn	Cd	Cu	Zn	Pb	Cr	Ni
Chirimba at Cori	3.185 ±0.931	0.018 ±0.135	0.563 ±0.067	2.828 ±0.231	ND	0.087 ±0.056	0.443 ±0.088	3.351 ±0.541	0.291 ±0.013	0.313 ±0.065	3.035 ±0.125	0.194 ±0.068	0.013 ±0.001	0.028 ±0.013
Chirimba at Machinjiri	1.903 ±0.284	0.029 ±0.013	0.129 ±0.009	1.393 ±.333	0.132 ±0.031	0.057 ±0.023	0.019 ±0.071	0.522 ±0.123	0.054 ±0.001	0.154 ±0.124	0.351 ±0.123	0.121 ±0.063	0.036 ±0.011	0.416 ±0.032
Mudi at MDI	5.641 ±0.963	ND	0.091 ±0.035	1.16 ± 0.611	0.198 ±0.132	0.036 ±0.002	0.146 ±0.003	4.875 ±1.112	0.362 ±0.041	0.171 ±0.023	1.258 ±0.047	0.704 ±0.126	0.045 ±0.016	ND
Mudi at SRN	1.782 ±0.491	ND	0.223 ±0.029	2.734 ±0.328	0.266 ±0.204	0.335 ±0.057	ND	13.521 ±1.088	0.171 ±0.008	0.105 ±0.041	2.263 ±1.334	0.972 ±0.012	0.663 ±0.031	0.233 ±0.036
Soche WWTP raw	1.438 ±0.196	ND	0.374 ±0.064	2.993 ±0.640	0.174 ±0.100	ND	ND	4.203 ±0.805	0.836 ±0.078	0.196 ±0.031	2.603 ±0.072	0.782 ±0.013	0.024 ±0.007	0.073 ±0.012
Soche WWTP effluent	0.586 ±0.168	ND	0.299 ±0.052	3.270 ±0.149	ND	ND	ND	0.281 ±0.142	0.142 ±0.031	0.175 ±0.136	2.621 ±1.209	0.042 ±0.001	ND	ND
Blantyre WWTP raw	0.432 ±0.075	ND	0.826 ±0.237	2.018 ±0.512	ND	0.431 ±0.137	ND	3.862 ±0.335	0.912 ±0.012	1.804 ±0.201	6.188 ±0.527	0.186 ±0.093	0.514 ±0.003	0.025 ±0.002
Blantyre WWTP effluent	0.731 ±0.406	ND	0.265 ±0.039	1.241 ±0.222	ND	0.029 ±0.003	ND	7.393 ±2.654	0.044 ±0.018	0.614 ±0.335	2.149 ±0.893	0.224 ±0.016	0.035 ±0.019	0.016 ±0.014
Nasolo at BNC	1.725 ±0.533	0.024 ±0.403	0.113 ±0.012	2.289 ±0.472	0.702 ±0.076	0.028 ±0.001	ND	4.213 ±1.018	0.468 ±0.031	0.056 ±1.032	4.426 ±1.244	0.423 ±0.072	0.075 ±0.041	ND
Nasolo at SRN	2.333 ±1.452	0.035 ±1.062	0.122 ±0.095	1.915 ±0.707	0.965 ±0.076	ND	ND	5.061 ±0.198	0.022 ±0.142	1.498 ±0.417	1.744 ±0.124	0.071 ±0.031	ND	0.061 ±0.012
Michiru	3.817 ±0.601	ND	0.002 ±0.012	0.202 ±0.159	ND	ND	ND	2.399 ±0.544	0.393 ±0.017	0.605 ±0.386	0.203 ±0.091	ND	ND	ND
Mangunda	3.968 ±1.098	0.022 ±1.463	0.026 ± 0.02	0.594 ±0.595	0.523 ±0.005	ND	ND	12.421 ±1.711	0.586 ±0.047	0.326 ±0.026	0.922 ±0.024	0.376 ±0.012	0.153 ±0.055	0.421 ±0.026
Limbe WWTP effluent	0.860 ±0.456	ND	0.229 ± 0.07	2.855 ±0.039	0.141 ±0.016	ND	ND	0.793 ± .117	0.796 ±0.141	0.223 ±0.041	0.726 ±0.124	ND	0.043 ±0.012	0.051 ±0.013
Limbe WWTP raw	2.769 ±1.586	ND	0.092 ±0.016	0.936 ±0.678	0.162 ±0.102	ND	ND	2.065 ±0.408	0.082 ±0.013	2.302 ±0.135	0.459 ±0.073	0.323 ±0.094	0.061 ±0.004	0.035 ±0.004
Limbe at Mpingwe	3.599 ±1.586	ND	0.029 ± 0.02	0.923 ±0.767	ND	ND	ND	16.132 ±1.527	0.074 ±0.034	0.551 ±0.310	0.496 ±0.043	0.406 ±0.072	ND	ND
Limbe at Highway	3.950 ±0.998	ND	0.09 ± 0.036	2.436 ±0.378	0.351 ±0.076	ND	ND	4.405 ±1.203	0.428 ±0.153	0.123 ±0.052	1.751 ±0.381	0.475 ±0.024	ND	0.063 ±0.013
Naperi at Rainbow	4.634 ±1.289	0.016 ±0.217	0.225 ± 0.04	1.774 ±0.308	0.263 ± 0.132	ND	ND	3.401 ± 0.467	0.039 ± 0.019	0.596 ± 0.102	5.358 ± 1.134	0.461 ± 0.068	0.011 ± 0.002	ND
Naperi at Moi	1.913 ± 0.2	ND	0.077 ±0.037	0.230 ±0.005	0.14 1±0.011	0.016 ±0.007	ND	7.164 ±0.842	0.116 ±0.042	1.418 ±0.426	0.988 ±0.403	0.037 ±0.008	0.063 ±0.014	ND

ND – not detected (below detection limit)

Table 2. Levels of heavy metals in *S. aequinoctialis* (in mg/kg dry weight)

Site	Dry season (mg/kg dw; $x \pm$ SD)									Rainy season (mg/kg dw; $x \pm$ SD)								
	pH	%OM	Mn	Ni	Cr	Cd	Pb	Cu	Zn	pH	%OM	Mn	Ni	Cr	Cd	Pb	Cu	Zn
Chirimba (A)	7.3	1.971 ±0.313	12.40 ±0.88	0.64 ±0.07	0.49 ±0.02	0.01 ±1.05	1.38 ±0.25	0.74 ±0.14	3.15 ±0.44	7.93	3.101 ±0.102	17.89 ± 0.14	0.001 ±0.001	1.24 ±0.35	0.04 ±0.05	0.55 ±.61	1.69 ±0.14	5.03 ±.24
Chirimba (B)	6.27	4.789 ±0.303	10.11 ±0.33	0.44 ±.23	0.15 ±1.06	0.02 ±0.04	2.48 ±0.31	1.63 ±0.04	3.17 ±0.49	6.94	4.759 ±0.327	15.70 ± 0.54	1.52 ± 0.18	1.28 ±0.29	ND	1.42 ±.27	1.62 ±0.40	9.54 0.51
Mudi (A)	7.22	5.674 ±0.379	14.81 ±0.91	2.94 ±0.86	5.62 ±1.09	0.16 ±0.04	3.49 ±0.67	1.31 ±0.16	5.22 ±0.53	6.73	4.455 ±1.764	13.98 ± 0.13	2.53 ± 0.18	4.42 ±0.13	ND	2.44± 0.41	3.35 ±0.84	13.94 ±0.83
Mudi (B)	6.62	0.880 ±0.658	14.60 ±0.82	0.05 ±0.01	0.93 ±0.06	0.07 ±1.01	1.75 ±0.44	7.31 ±2.32	0.26 ±0.20	6.95	0.588 ±0.428	10.26 ± 1.03	0.93 ± 0.34	2.37 ±0.92	ND	2.95 ±0.88	0.90 ±0.05	5.17 ±0.37
Michiru stream	6.63	2.732 ±0.668	14.40 ±0.32	1.45 ±0.37	1.09 ±0.15	ND	0.03 ±0.18	2.47 ±0.45	3.27 ±1.00	6.47	2.727 ±0.680	10.31 ± 2.92	2.89 ± 0.52	2.60 0.56	ND	2.32 ±0.37	0.41 ±0.22	3.31 ±0.30
Naperi (A)	6.52	4.021 ±0.392	12.85 ±0.86	0.86 ±0.11	2.81 ±0.29	0.03 ±1.01	0.22 ±0.09	7.11 ±1.42	2.16 ±0.59	7.03	4.082 ±1.184	13.73 ±0.85	2.46 ± 0.65	2.85 ±1.24	ND	1.84 ±0.26	1.48 ±0.36	5.52 ±1.09
Naperi (B)	6.67	0.559 ±0.304	13.34 ±0.34	4.32 ±.56	1.35 ±0.30	ND	2.26 ±0.31	3.28 ±0.97	1.57 ±0.50	7.26	3.921 ±0.302	14.15 ±0.41	1.09 ± 0.25	1.80 0.42	ND	1.77 ±0.19	1.90 ±0.16	6.29 ±0.19
Blantyre WWTP	6.33	7.755 ±0.778	27.43 ±0.95	2.54 ±0.46	8.19 ±0.70	0.13± 0.05	3.04 ±0.09	0.90 ±0.05	3.22± 0.52	6.47	9.266 ± .404	14.18 ±0.50	1.15 ± 0.19	6.83 ±1.67	ND	2.46 ±0.23	5.87 ±0.86	17.45 ±0.92
Nasolo s(A)	6.89	3.310 ±0.042	16.87 ±0.71	0.07 ±.12	1.99 ±1.51	ND	0.31 ±0.25	1.71 ±0.47	6.26 ±1.30	7.21	2.460 ±0.858	10.64 ±1.050	0.41 ± 0.13	0.53 ±0.16	ND	1.97 ±0.20	0.75 ±0.17	7.65 ±1.84
Nasolo (B)	6.94	2.691 ± .445	12.01 ±2.15	0.97 ±.14	0.05 ±0.03	ND	0.04 ±0.15	0.31 ±0.03	7.04 ±.56	7.22	2.923 ± .806	11.31 ± 1.86	0.93 ± 0.12	0.54 0.47	ND	1.61 0.31	1.19 ±.15	11.09 0.21
Namangunda	6.37	9.357 ±0.525	9.00 ±0.18	4.14 ±0.37	3.48 ±0.58	0.13 ±2.03	0.71 ±0.09	10.13 ±0.98	1.44 ±.55	6.47	1.729 ± .228	10.96 ± 0.44	ND	ND	ND	0.51 ±.14	0.13 0.03	1.37 ±.09
Soche WWTP	6.31	5.834 ±0.179	26.78 ±1.89	1.19± 0.07	1.91 ±0.81	0.04 ±0.02	1.17 ±0.18	3.02 ±0.40	14.46 ±0.64	6.52	6.702 ±0.807	14.29 ± 0.23	0.80 ± 0.17	1.63 ±0.22	ND	1.47 0.034	3.18 ±0.13	16.18 ±0.45
Limbe WWTP	6.68	0.582 ±0.203	31.43 ±0.36	0.56 ±0.14	0.21 ±0.09	0.18 ±0.02	0.41 ±0.20	0.55 ±0.07	2.15 ±0.02	7.56	1.604 ±0.050	14.50 ±0.14	1.40 ±0.11	1.19 ±0.16	ND	1.16 ±0.23	1.67 ±0.15	3.83 ±0.64
Limbe stream (A)	6.5	1.866 ±0.134	17.43 ±0.23	1.82 ±0.18	0.07 ±0.15	ND	0.64± 0.08	0.12 ±1.14	2.58 ±0.15	6.93	0.873 ±0.832	10.84 ±1.01	0.26 ± 0.06	0.37 ±0.54	ND	0.82 ±0.05	0.35 ±0.06	2.41 ±0.30
Limbe stream (B)	7.06	2.019 ±0.286	25.96 ±1.41	0.03 ±0.01	0.27 ±0.06	0.02± 0.01	1.15 ±.12	1.04 ±0.01	6.05 ±0.05	7.82	2.37 ± 0.731	12.24 ±0.33	0.59 ± 0.16	1.37 ±0.72	ND	2.06 ±0.39	1.16 ±0.22	7.61± 0.52

x = mean value (n = 3); SD = standard deviation; ND = not detected (below detection limit)

Table 3. Metal concentrations in stream sediments soils and soils around WWTPs in the study area

Site	Dry season (mg/kg dw; $x \pm SD$)							Rainy season (mg/kg dw; $x \pm SD$)						
	Mn	Ni	Cr	Cd	Pb	Cu	Zn	Mn	Ni	Cr	Cd	Pb	Cu	Zn
Chirimba stream (A)	3.31 ±1.50	0.45 ±0.04	ND	0.17 ± 0.04	0.42 0.15	0.07 ± 0.06	4.30 ±0.08	5.24 ± 2.12	0.64 ±0.12	ND	0.123 ± 0.012	0.14 0.32	0.17 ± 0.16	3.67 ± 1.32
Chirimba stream (B)	4.66 ± 0.75	0.93 ±0.02	ND	0.29 ± 0.01	0.27 0.02	0.14 ± 0.04	1.26 ±1.01	1.01 ± 0.13	0.30 ±0.09	ND	0.128 ± 0.016	0.12 0.50	ND	3.43 ± 0.27
Mudi stream (A)	4.29 ±2.50	0.74 ±0.03	0.018 ±0.003	0.43 ± 1.03	0.04 ±0.03	0.35 ± 0.03	0.70 ±0.26	2.36 ± 0.28	0.38 ±0.06	ND	0.144 ± 0.002	0.34 ±0.44	0.14 ± 0.07	4.08 ± 0.77
Mudi stream (B)	4.03 ± 1.69	0.13 ±0.01	ND	0.17 ± 0.01	0.19 ±0.01	0.24 ± 0.09	2.13 ± 0.11	6.39 ± 0.52	0.53 ± 0.17	ND	0.136 ± 0.007	ND	0.19 ± 0.05	3.71 0.31
Michiru stream	2.73 ±0.48	0.04 ±0.02	ND	0.14 ± 2.05	ND	0.005 ± 0.002	1.82 ± 0.14	1.94 ± 0.08	0.37 ± 0.02	ND	0.115 ± 0.004	ND	ND	3.00 ± 0.17
Naperi stream (A)	1.60 ±0.72	0.35 ±0.05	0.008 ± 0.001	0.12 ± 0.55	0.03 ± 0.11	0.128 ± 0.013	0.70 ± 0.27	3.69 ± 0.38	0.38 ± 0.02	ND	0.108 ± 0.014	ND	0.004 ± 0.001	4.03 ± 0.26
Naperi stream (B)	1.88 ±0.85	0.26 ±0.04	ND	0.22 ± 0.49	ND	0.099 ± 0.027	0.74 ± 0.20	3.16 ± 1.752	0.33 ± 0.08	ND	0.123 ± 0.013	ND	0.040 ± 0.009	3.16 ± 2.385
Blantyre WWTP	2.08 ±0.34	0.54 ±0.05	0.029 ± 0.012	0.17 ± 0.02	0.08 ± 0.39	0.202 ± 0.002	4.29 ± 0.11	3.25 ± 0.25	0.32 ± 0.04	ND	0.118 ± 0.006	ND	0.41 ± 0.22	4.82 ± 0.74
Nasolo stream (A)	4.02 ±0.27	0.25 ±0.04	ND	0.19 ± 0.02	0.45 ± 0.14	0.373 ± 0.065	1.90 ± 1.05	3.01 ± 0.68	0.30 ± 0.06	ND	0.119 ± 0.002	0.80 ± 0.18	0.05 ± 0.01	4.35 ± 0.79
Nasolo stream (B)	1.66 ±0.29	0.65 ±0.03	ND	0.50 ± 0.01	0.18 ± 0.02	0.136 ± .004	1.15 ± .08	2.16 ± 0.13	0.33 ± .03	ND	0.129 ± 0.003	0.48 ± 0.33	0.03 ± 0.01	2.78 ± 0.24
Mzedi Stream	3.83 ±0.81	0.46 ±0.04	0.031 ± .018	0.24 ± 0.85	0.48 ± 0.06	0.281 ± .041	5.11 ± .92	3.90 ± 1.46	0.29 ± .03	ND	0.121 ± 0.004	ND	ND	0.66 ± 0.12
Soche WWTP	5.89 ± 0.18	0.36 ±0.04	0.014 ± .006	0.33 ±0.01	0.34 ± 0.05	0.044 ± .003	4.06 ± .04	4.39 ± 0.57	0.87 ± .44	ND	0.114 ± 0.010	ND	0.09 ± 0.07	5.27 ± 0.68
Limbe WWTP	7.58 ± 0.35	0.27 ±0.02	ND	0.55 ±0.02	0.14 ± 1.03	0.265 ± .119	1.40 ± .32	9.62 ± 0.49	0.50 ± .03	ND	0.127 ± 0.019	ND	0.21 ± 0.16	2.71 ± 0.26
Limbe stream (A)	3.86 ± 0.55	0.05 ±0.04	ND	0.12 ± 0.27	0.02 ± 0.11	0.028 ± .006	1.51 ± .58	7.98 ± 0.58	0.35 ± .03	ND	0.109 ± 0.013	ND	0.03 ± 0.01	2.78 ± 0.27
Limbe stream (B)	1.92 ± 0.23	0.84 ±0.05	ND	0.19 ± 0.12	0.3 ± 1.09	0.155 ± .012	4.45 ± .44	4.19 ± 1.50	0.28 ± .07	ND	0.114 ± 0.009	ND	0.11 ± 0.04	3.98 ± 1.66

Table 4. Metal concentrations in *A. icteria* inhabiting stream- sediments and soils around WWTPs in Blantyre City, Malawi

	%OM	pH	[Cd]$_s$	[Cd]$_w$	[Pb]$_s$	[Pb]$_w$	[Mn]$_s$	[Mn]$_w$	[Cu]$_s$	[Cu]$_w$	[Zn]$_s$	[Zn]$_w$	[Cr]$_s$	[Cr]$_w$	[Ni]$_s$	[Ni]$_w$
%OM	1	-0.39	ND	-0.09	0.23	-0.68	0.49	-0.46	0.89**	0.49	0.84**	0.61*	0.69**	ND	0.17	0.17
pH	-0.40	1	ND	0.03	-0.25	-0.16	0.36	0.33	-0.25	-0.21	-0.28	0.02	-0.53*	ND	-0.51	0.016
[Cd]$_s$	0.25	-0.02	1	ND	ND	ND	ND	ND	ND	ND	ND	ND	ND	ND	ND	ND
[Cd]$_w$	-0.03	0.18	0.63*	1	0.29	-0.29	0.12	-0.11	0.15	0.25	0.21	-0.08	0.16	ND	0.16	-004
[Pb]$_s$	0.28	0.02	0.20	0.04	1	0.54	-0.32	-0.31	0.33	0.31	0.41	0.53*	0.59*	ND	0.46	-0.15
[Pb]$_w$	0.18	0.10	-0.31	-0.097	-0.23	1	-0.91*	-0.045	0.45	-0.82	0.07	0.35	-0.28	ND	-0.19	-0.50
[Mn]$_s$	-0.08	-0.09	0.31	0.25	0.03	-0.20	1	-0.012	0.50	0.31	0.31	0.31	0.16	ND	-0.08	0.38
[Mn]$_w$	-0.02	-0.17	0.39	0.50	0.000	0.13	0.39	1	-0.17	0.15	-0.41	-0.16	-0.26	ND	0.39	0.35
[Cu]$_s$	0.35	-0.38	-0.02	-0.30	-0.13	0.23	-0.41	-0.09	1	0.73**	0.85**	0.63*	0.81**	ND	0.09	0.21
[Cu]$_w$	0.26	0.22	0.83**	0.33	0.24	0.12	0.05	0.26	0.18	1	0.42	0.31	0.72**	ND	0.007	0.092
[Zn]$_s$	0.21	0.03	-0.23	0.29	-0.08	0.25	0.36	0.19	-0.23	-0.15	1	0.72**	0.56*	ND	0.04	0.23
[Zn]$_w$	0.47	-0.07	-0.15	-0.25	0.026	0.61*	0.27	-0.007	-0.31	-0.04	0.22	1	0.48	ND	-0.08	0.41
[Cr]$_s$	0.71**	-0.15	0.53	-0.09	0.56*	-0.24	0.154	-0.17	0.18	0.41	-0.007	0.22	1	ND	0.46	-0.09
[Cr]$_w$	0.96**	-0.23	0.73	0.011	0.33	0.44	0.004	-0.03	0.15	0.50	-0.38	0.73	0.56	1	ND	ND
[Ni]$_s$	0.44	-0.20	0.60	-0.003	0.36	-0.08	-0.236	-0.19	0.09	0.04	-0.21	0.05	0.49	0.86	1	-0.21
[Ni]$_w$	0.36	0.20	-0.20	0.34	0.46	0.099	-0.04	-0.13	0.28	0.24	0.22	0.16	0.17	0.33	-0.06	1

** Correlation is significant at the 0.01 level; * Correlation is significant at the 0.05 level; ND = not determined

Table 5. Pearson's correlation coefficients of soil OM, pH-H$_2$O, metal soil and *A.icteria* metal content (Upper panel – rainy season Lower panel – dry season)

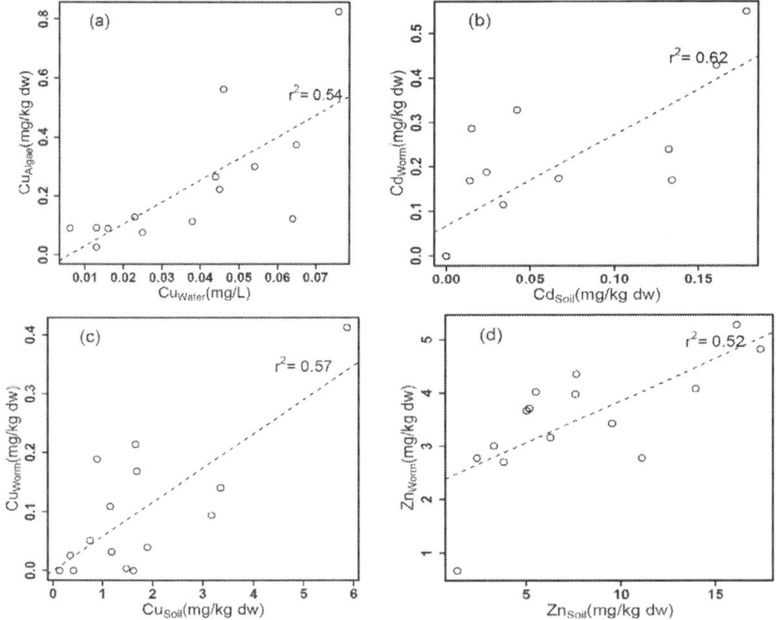

Fig. 2. Scatter plots showing relationship between metal content in *A.icteria* and sediments (*b –d*) and metal content in *S. aequinoctialis* and stream water (*a*). Dry season data is used for plot (b) and rainy season data is used for the other plots. No significant correlations were found for the other combinations of metals.

Site	Dry season							Rainy season						
	Mn	Ni	Cr	Cd	Pb	Cu	Zn	Mn	Ni	Cr	Cd	Pb	Cu	Zn
Chirimba at Cori	ND	0.07	ND	3.99	5.24	ND	6.05	91.00	1.05	ND	0.49	0.00	12.24	9.59
Chirimba at Machinjiri	ND	1.06	ND	0.74	3.46	ND	0.63	12.28	0.05	ND	0.71	1.22	5.61	9.41
Mudi at MDI	81.25	ND	ND	3.26	18.53	ND	0.84	23.12	0.42	ND	0.00	2.51	15.17	10.00
Mudi at SRN	ND	0.67	ND	2.01	15.19	ND	0.87	10.01	ND	0.85	0.00	2.92	4.96	26.80
Soche WWTP raw	ND	0.19	ND	9.39	18.62	ND	3.70	3.94	ND	ND	0.00	1.58	5.75	12.85
Soche WWTP effluent	ND	ND	ND	1.63	0.72	ND	3.69	1.53	ND	ND	0.00	0.00	5.54	16.77
Blantyre WWTP raw	ND	0.06	ND	11.26	3.96	ND	9.18	0.99	ND	1.45	0.00	0.00	10.87	11.66
Blantyre WWTP effluent	ND	0.04	ND	0.54	6.59	ND	2.90	1.61	ND	2.07	0.00	0.00	6.02	9.19
Nasolo at BNC	ND	0.00	ND	5.71	6.13	ND	4.10	4.11	ND	1.12	0.30	7.63	2.97	17.21
Nasolo at SRN	ND	0.14	ND	0.28	0.96	ND	1.83	5.11	ND	ND	0.36	20.10	1.91	12.04
Michiru	ND	ND	ND	4.57	0.00	ND	0.39	68.16	ND	ND	0.00	ND	ND	1.45
Mangunda	23.44	1.07	ND	6.51	3.84	ND	1.83	8.11	ND	ND	0.22	5.13	2.00	3.93
Limbe WWTP effluent	ND	0.15	ND	10.34	0.00	ND	1.08	1.68	ND	ND	0.00	ND	ND	9.81
Limbe WWTP raw	ND	0.16	ND	1.03	4.97	ND	0.73	4.42	ND	ND	0.00	4.05	7.08	3.55
Limbe at Mpingwe	ND	0.00	ND	1.06	12.30	ND	0.88	73.45	ND	ND	0.00	0.00	ND	2.29
Limbe at Highway	ND	0.15	ND	5.94	5.34	ND	2.77	23.51	ND	ND	0.00	4.23	5.63	14.16
Naperi at Rainbow	ND	0.00	ND	0.48	41.91	ND	7.50	7.92	ND	ND	0.18	6.74	ND	12.41
Naperi at Moi	ND	0.00	ND	1.41	0.97	ND	1.59	4.12	ND	ND	0.00	0.00	3.08	1.93

ND = not determined (because metal content was below detection limit in either water or algae)

Table 6. Bioconcentration factors (BCF) for rainy and dry seasons for *S. aequinoctialis* from the sampled streams in Blantyre City, Malawi.

Site	Dry season							Rainy season						
	Mn	Ni	Cr	Cd	Pb	Cu	Zn	Mn	Ni	Cr	Cd	Pb	Cu	Zn
Chirimba stream (A)	0.27	0.70	ND	12.07	0.31	0.10	1.36	0.29	ND	ND	2.98	0.24	0.100	0.73
Chirimba stream (B)	0.46	2.10	ND	19.13	0.11	0.09	0.40	0.06	0.19	ND	ND	0.08	ND	0.36
Mudi stream (A)	0.29	0.25	0.0032	2.66	0.01	0.27	0.13	0.17	0.15	ND	ND	0.14	0.042	0.29
Mudi stream (B)	0.28	2.46	ND	2.60	0.11	0.03	8.35	0.62	0.57	ND	ND	ND	0.211	0.72
Michiru stream	0.19	0.03	ND	ND	ND	0.00	0.56	0.19	0.13	ND	ND	ND	ND	0.91
Naperi stream (A)	0.12	0.41	0.0028	3.38	0.12	0.02	0.32	0.27	0.15	ND	ND	ND	0.003	0.73
Naperi stream (B)	0.14	0.06	ND	ND	ND	0.03	0.47	0.22	0.31	ND	ND	ND	0.021	0.50
Blantyre WWTP	0.08	0.21	0.0035	1.27	0.03	0.22	1.33	0.23	0.28	ND	ND	ND	0.070	0.28
Nasolo stream (A)	0.24	3.36	ND	ND	1.44	0.22	0.30	0.28	0.72	ND	ND	0.41	0.068	0.57
Nasolo stream (B)	0.14	0.67	ND	ND	4.44	0.43	0.16	0.19	0.35	ND	ND	0.30	0.027	0.25
Mzedi Stream	0.43	0.11	0.0089	1.84	0.67	0.03	3.56	0.36	ND	ND	ND	ND	ND	0.48
Soche WWTP	0.22	0.30	0.0073	7.83	0.29	0.01	0.28	0.31	1.08	ND	ND	ND	0.030	0.33
Limbe WWTP	0.24	0.49	ND	3.08	0.33	0.48	0.65	0.66	0.35	ND	ND	ND	0.128	0.71
Limbe stream (A)	0.22	0.03	ND	ND	0.03	0.24	0.58	0.74	1.34	ND	ND	ND	0.074	1.15
Limbe stream (B)	0.07	32.38	ND	7.83	0.26	0.15	0.74	0.34	0.48	ND	ND	ND	0.094	0.52

ND = not determined

Table 7. Bioconcentration factors (BCF) for rainy and dry seasons for *A.icteria* from the sampled stream-bank soils and soils around WWTPs in Blantyre City, Malawi.

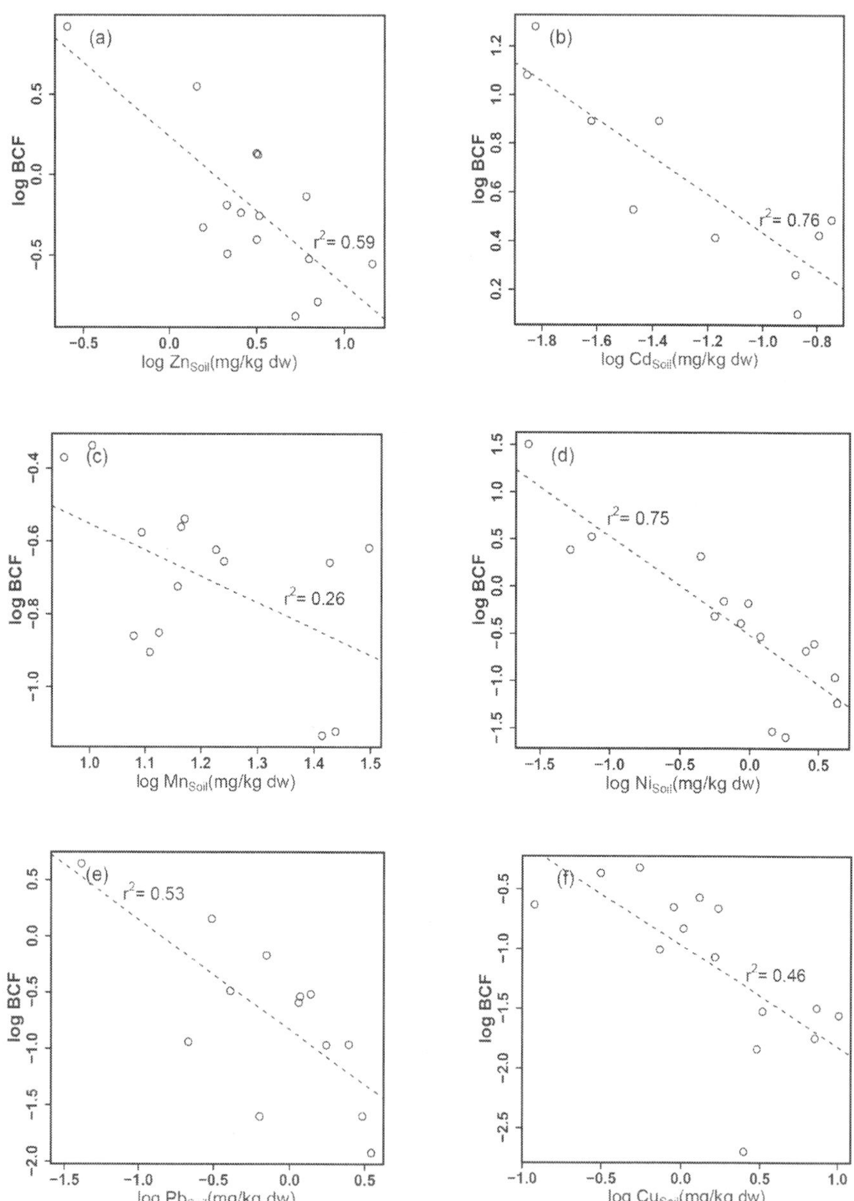

Fig. 3. Plots of log BCF against metal concentration in the soil using dry season data.

References		Species	Equation #	Cd		Pb		Cu		Zn	
				a	*b*	*a*	*b*	*a*	*b*	*a*	*b*
Heikens et al. (2001)	Bibliographic study	Lumbricidae (mixed)	1	0.39	1.1	0.62	-0.3	0.17	1.8	0.17	1.8
Ma et al. (2004)	Field data	Lumbricidae (mixed)	2	0.556	1.39	0.556	0.626	0.327	0.776	0.212	2.49
Neuhauser eta al. (1995)	Bibliographic studies	Lumbricidae (all species mixed)	3	0.66	1.21	0.74	0.05	0.27	2.09	0.27	2.09
Wright and Stringer (1980)	Soil quantity: field study	*Aporrectodea caliginosa*	4	0.32	0.33	0.9	-0.8	0.01	0.23	0.01	0.23
		Aporrectodea longa	5	0.3	-0.3	0.5	-0.1	0.1	2.5	0.1	2.5
		Aporrectodea rosea	6	0.5	0.7	0.5	-0.1	0.5	1.1	0.5	1.1

Table 8. Regression equations, Log M_{ew} = alogM_s + b for Lumbricidae, *Aporrectodea caliginosa, Aporrectodea longa, Aporrectodea rosea* and Cd, Cu, Pb and Zn from literature

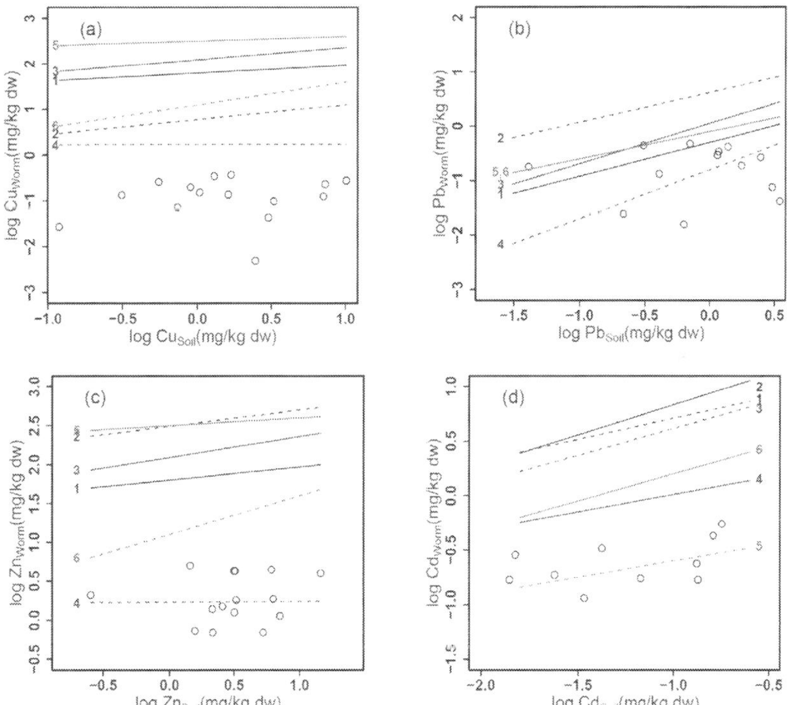

Fig. 4. Relationships between the internal Cd, Cu, Pb, Zn in earthworms and the total soil concentration, in comparison with regressions from the literature. The number on the lines corresponds to the number of regression model in Table 8, calculated using the soil data from this study. The dots are data from this study (*A.icteria*) for dry season.

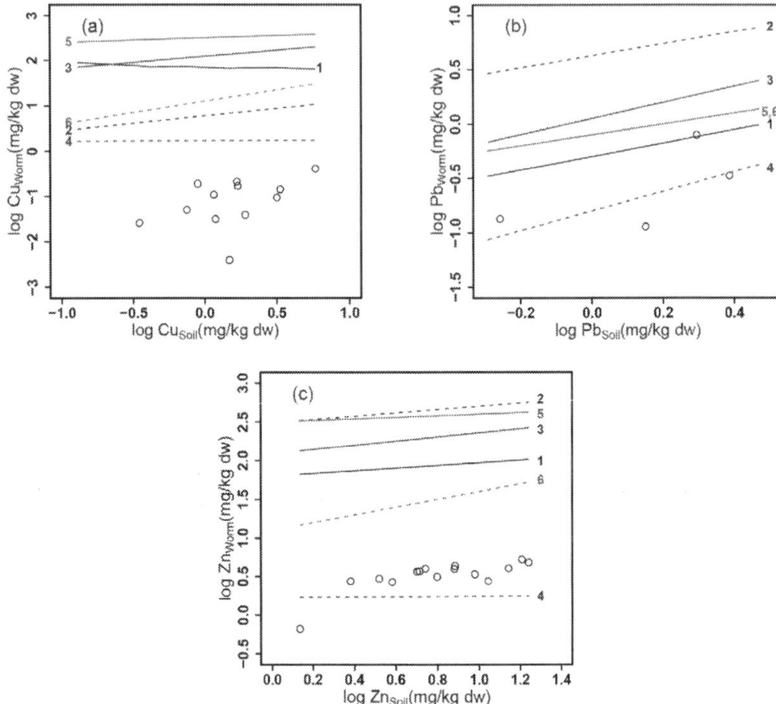

Fig. 5. Relationships between the internal Cd, Cu, Pb, Zn in earthworms and the total soil concentration, in comparison with regressions from the literature. The number on the lines corresponds to the number of regression model in Table 8, calculated using the soil data from this study. The dots are data from this study (*A.icteria*) for wet season.

Kamitani and Kaneko, 2007; Hsu et al., 2006; Dai et al, 2004; Ireland, 1983; Ma, 1982). The solubility of heavy metals in soil (pore) water is important for bioaccumulation by earthworms as the main pathways for chemical absorption are the skin (for soluble elements), gut transit and digestion (Weltje, 1998). The BCF order also reflects the affinity order for the specific adsorption of metal cations in soil: Pb > Cu > Zn > Cd. Cadmium tends to be more mobile in soils and therefore more available to earthworms than other heavy metals (Ma, 2004). Fig. 3 shows decreasing BCF with soil concentration of the metals indicating that bioconcentration depends on the metal concentrations in the soil and is greater at low soil concentrations. This implies that *A. icteria* exhibits metal regulation at high exposure rates (Neuhauser et al., 1995).

There is a significant consensus in the literature that bioaccumulation of heavy metals by earthworms is dependent on earthworm species and type of metal (see e.g. van Vliet et al., 2005, 2006; Vijver, 2007; Kamitani and Kaneko, 2007; Ernst et al., 2008).

Hence, the accumulation patterns of Cd, Cu, Zn and Pb in *A. icteria* (this study) were compared to those of other lumbricid earthworms, using the regression models from literature (Heikens et al., 2001; Neuhauser et al., 1995; Ma, 2004; Wright and Stringer, 1980) (see Table 8). Data for *A. icteria* is not readily available in the literature for comparison with our study. Figs 4 and 5 show that *A. icteria* (present study) accumulated relatively less Cu than the other species at the same soil concentration. However, similar accumulation levels were observed for Pb and Zn with *A. caliginosa* and for Cd with *A. longa* (Figs 4 and 5), even at our very low exposure levels. For all species, there is a metal dependent increase in body concentration with increasing soil concentration, following the order: Pb > Cd > Cu = Zn (Figs. 4 and 5). Cu and Zn are essential metals and are physiologically regulated by earthworms, resulting in a relatively constant body concentration with respect to soil metal concentrations (Panda et al., 1999; Heikens, et al., 2001; Morgan and Morgan, 1988; Lukkari, 2004). In contrast, Cd and Pb are non-essential metals and are not regulated resulting in metal increase with increasing soil concentrations (Spurgeon and Hopkin, 1999).

The plots in Figs. 4 and 5 also show that the regression models generated for all species of lumbricid earthworms would overestimate the amount of heavy metals in *A. icteria* living in the stream bank soils and soils around the WWTPs in Blantyre City. This result supports the general consensus that the degrees of heavy metal accumulation in earthworms show different, metal- and species-specific patterns. The observed differences in accumulation patterns are usually attributed to differences in metal kinetics of the earthworms, exposure route and food preference. In addition, it seems that the accuracy of the regression models is lost when they are generalised at family level but may be possible to relate accumulation patterns within the same genera. However, it should be noted that the comparison should be made with caution because metal availability is dependent on several environmental factors such as soil pH, cation exchange capacity, OM and Ca^{2+} (Christensen, 1989, Ma, 1982; Corp and Morgan, 1991; Peijnenburg et al., 1999a, 1999b). These factors may differ between studied field soils.

5. Conclusions

The study obtained concentrations of manganese, cadmium and lead in periphyton (*S. aequinoctialis*) in higher levels than in the corresponding water, implying that *S. aequinoctialis* accumulates these heavy metals. The results indicate the potential of periphyton as a biological indicator of heavy metal pollution. Heavy metals concentrations therefore measured in macroalgae species can give a picture of the quality of our surrounding environment. In addition, the levels of most of the heavy metals were higher than drinking water standards. It was also found that the general trend was that of high heavy metal values for water samples in the dry season than in the rainy season. The relatively low heavy metal levels in the rainy season were attributed to dilution. The study also showed that *A. icteria* can accumulate cadmium, but not lead and manganese. This work supports the published results that metal- and species-specific accumulation patterns of non-essential heavy metals in earthworms occur. In addition, the work extends the database to even very low exposure levels and therefore generates more information for *A. icteria*. Metal accumulation shows seasonal variations with significant correlations and multiple regression models between soil and internal metal content being more apparent in the rainy season. Similar accumulation levels observed between *A. icteria* and *A. caliginosa* for Pb and Zn and *A. longa* for Cd point to a possibility to relate accumulation patterns within the same genera, albeit metal specific. Further, it seems that the accuracy of regression models is lost when they are generalised at family level, making generalisations at that level difficult.

6. Acknowledgements

The authors would like to thank University of Malawi, The Polytechnic for funding this study. They are greatly indebted to Blantyre City Assembly for allowing them to collect samples in places that they use for their routine monitoring work. Thanks should also go to technical staff of Chancellor College Faculty of Science, all laboratory staff at Geological Surveys Department and National Herbarium and Botanical Gardens for their technical support.

7. References

Ackcali, I., Kucuksezgin, F. (2011). A biomonitoring study: Heavy metals in macroalgae from eastern Aegean coastal areas. *Marine Pollution Bulletin,* Vol 62, pp: 637-645, ISSN 0025326X

Alloway, B.J., & Ayres D.C. (1997). *Chemical Principles of Environmental Pollution* (2nd Edition), Blackie Academic & Professional, ISBN 0751403806, London

American Public Health Association, (APHA). (2005). *Standard Methods for the Examination of Water and Wastewater* (21st Edition), APHA, ISBN 0875532233, Washington DC

Anderson, J.M., & Ingram, J.S.I. (1993). *Tropical Soil Biology: A Handbook of Methods* (2nd Edition), CAB International, ISBN 0851988210, UK

Association of Official Analytical Chemists, (AOAC). (1990). *Official Methods of Analyses* (15th Edition), AOAC, ISBN 8787200848, New York, USA

Bamgbose, O., Odukoya, O., & Arowolo, T.O.A. (2000). Earthworms as bio-indicators of heavy metal pollution in dumpsites of Abeokuta city, Nigeria. *Rev. Biol. Trop.,* Vol. 48, No. 1, (March 2000), pp. 229 – 234,   ISSN 0034-7744

Bohn, H.L., McNeal, B.L. & O'Connor, G.A. (1985). *Soil Chemistry* (2nd Edition). John Wiley and Sons, ISBN 0471822175, New York

Burns, A., & Ryder, D.S. (2001). Potential for biofilms as biological indicators in Australian riverine systems. *Ecological Management & Restoration,* Vol. 2, No. 1, (April 2001), pp. 53–64, ISSN 14428903

Christensen, T.H. (1989). Cadmium soil sorption at low concentrations. Correlation woth soil parameters. *Water Air Soil Pollution,* Vol 44, pp: 71-82, 00496979

Chukwuma, C. (1998). Environmental issues and our chemical world – The need for a multidimensional approach in environmental safety, health and management. *Environmental Management and Health,* Vol. 9, No. 3, pp. (136- 143), ISSN 09566163

Corp, N., Morgan, A.J. (1991). Accumulation of heavy metals from polluted soils by the earthworm, *Lumbricus rubellus*: can laboratory exposure of 'control' worms reduce biomonitoring problems? *Environmental Pollution,* Vol 74, pp: 39-52, ISSN 02697491

Cuffney, T.F.,Meador, M.R., Porter, S.D. & Curtz, M.E. (2000). Responses of physical, chemical and biological indicators of water quality to a gradient of agricultural land use in the Yakima River Basin, Washington. *Environmental Monitoring and Assessment,* Vol. 64, No.1, (September 2000), pp. 259-270, ISSN 01676369

Dai, J., Becquer, T., Rouiller, J.H., Reversata, G., Bernhard-Reversata, F., Nahmania, J., Lavelle, P. (2004). Heavy metal accumulation by two earthworm species and its relationship to total and DTPA-extractable metals in soils. *Soil Biology and Biochemistry,* Vol 36, pp: 91–98, ISSN 00380717

DARES (Diatoms for Assessing River Ecological Status). (2004). Sampling protocol version 1.0., In: *Diatoms for Assessing River Ecological Status*, Accessed on 11/07/2010, Available from <http://craticula.ncl.ac.uk/DARES/methods/DARESProtocol DiatomSampling.pdf>

Davies, T.A., Volesky, B., Mucci, A. (2003). A review of the biochemistry of heavy metal biosorption by brown algae. *Water Research*. Vol 37, pp: 4311-4330, ISSN 00431354

EMAN (Ecological Monitoring and Assessment Network). (2004). Worm species richness, In: *Environment Canada*, Accesed on 11/07/2010, Available from <http://www.emanrese.ca/eman/ecotools/protocols/terrestrial/earthworm/intr o.html>

Ernst, G., Zimmermann, S., Christie, P. & Frey, B. (2008). Mercury, cadmium and lead concentrations in different ecophysiological groups of earthworms in forest soils. *Environmental Pollution*, Vol. 156, No. 3, (April 2008), pp. 1304 – 13, ISSN 02697491

Favero, N., Cattalini, F., Bertaggia, D. & Albergoni, V. (1996). Metal accumulation in a biological indicator (*Ulva rigida*) from the lagoon of Venice (Italy). *Archives of Environmental Contamination and Toxicology*, Vol. 31, No. 1, (January 1996), pp.9-18, ISSN 14320703

Filho, A.G.M., Andrade, L.R., Karez, C.S., Farina, M., Pfeiffer, W.C. (1999). Brown algae species as bimonitors of Zn and Cd at Sepetiba Bay, Rio de Janeiro, Brazil. *Marine Environmental Research*. Vol 48, pp:213-224, ISSN 01411136

Gekeler, W., Grill, E., Winnacker, E.L. & Zenk, M.H. (1988). Algae sequester heavy metals via synthesis of phytochelatin complexes. *Archives of Microbiology*, Vol. 150, No. 2, pp.197-202, ISSN 1432072X

Georgescu, B. & Weber, C. (2007). The role of earthworms as biological indicators of soil contamination. *Animal Science and Biotechnologies*, Vol. 64, No. 1 2, ISSN 18435254

Garnharm, G.W., Codd, G.A., Gadd, G.M. (1992). Kinetics of uptake intracellular locations of cobalt, manganese and zinc in the estuarine green alga Chlorella salina. *Applied Microbiology and Biotechnology*, Vol 37, pp: 270–276, ISSN 14320614

Heikens, A., Peijnenburg, W.J.G.M., Hendriks, A.J. (2001). Bioaccumulation of heavy metals in terrestrial invertebrates. *Environmental Pollution*, Vol113, pp: 385-393, ISSN 02697491

Hill, B.H., Herlihy, A.T., Kaufmann, P.R., Stevenson, R.J., McCormick F.H. and Johnson C.B., (2000) Use of periphyton assemblage data as an index of biotic integrity. *Journal of the North American Benthological Society*, Vol. 19, No. 1, (March 2000), pp. 50-67, ISSN 08873593

Hoffman, S. (1996). My trip to Upper Klamath Lake during the 1996 full eclipsed harvest moon, Accessed on 21/06/2010, Available from <http://www.algae-world.com/algae/12.htmL>

Holan, Z.R., Volesky, B. & Prasetyo, I. (1993). Biosorption of cadmium by biomass of marine algae. *Biotechnology and Bioengineering*, (April 1993), Vol. 41, No. 8, pp. 819–825, ISSN 00063592

Hui, C. A.(2002). Lead distribution throughout soil, flora, and an invertebrate at a wetland skeet range. *Journal of Toxicology and Environmnetal Health*, Vol 65, pp: 1093–1107, ISSN 15287394

Hsu, M.J., Selvaraj, K., Agoramoorthy, G. (2006). Taiwan's industrial heavy metal pollution threatens terrestrial biota. *Environmental Pollution*, Vol 143, pp: 327–334, ISSN 02697491

Ireland, M.P. (1983). Heavy metal uptake and tissue distribution, In: *Earthworms Ecology from Darwin to Vermiculture*, E.J. Satchell, pp. 247 – 265, Chapman and Hall, ISBN 9788179102299, London

Ireland, M.P., Richards, K. S. (1977). The occurrence and localisation of heavy metals and glycogen in earthworms *Lumbricus rubellus* and *Dendrobaena rubida* from a heavy metal site. *Histochemistry*, Vol.51, pp: 153–166, ISSN 11214201

ISO (International Standards Organisation), 1994. Water quality- Determination of pH. ISO 10523-1, ISO, Geneva

Jin-fen, P., Rong-gen, L. & Li, M. (2000). A review of heavy metal adsorption by marine algae. *Chinese Journal of Oceanology and Limnology*, Vol. 18, No. 3, pp.260-264, ISSN 02544059

Kadewa, W. W., Henry, E. M., Masamba, W. R. M. & Kaunda, C. C. (2001). Impact of sewage sludge application to horticulture: A case of the city of Blantyre, *Proceedings of the first Chancellor College research dissemination conference*, Zomba, Malawi, (n.d)

Kamitani, T. & Kaneko, N. (2007). Species-specific heavy metal accumulation patterns of earthworms on afloodplain in Japan. *Ecotoxicology and Environmental Safety*, Vol. 66, No. 1, (January 2007), pp. 82-91, ISSN 01476513

Khan, I. S. A. N. (1990). Assessment of water pollution using diatom community structure and species distribution — A case study in a tropical river basin. *Internationale Revue der gesamten Hydrobiologie und Hydrographie*, Vol. 75, pp. 317–338, ISSN 15222632

Khosmanesh, A., Lawson, F. & Prince, I.G. (1996). Cadmium uptake by unicellular green microalgae. *The Chemical Engineering Journal and the Biochemical Engineering Journal*, Vol. 62, No.1, (April 1996), pp. 81-88, ISSN 09230467

Kuyeli, S.M. (2007). Assessment of industrial effluents and their impact on water quality in streams of Blantyre City, Malawi. Unpublished MSc Thesis. University of Malawi.

Lakudzala D.D, Tembo K.C and Manda I.K (1999). An investigation of chemical pollutants in Lower Shire River, Malawi. *Malawi Journal of Science and Technology*, Vol. 5, pp. 87-94, ISSN 1019 – 7079

Langdon, C.J., Pearce, T.G., Meharg, A.A., Semple, K.T. (2001) Resistance to copper toxicity in populations of the earthworms *Lumbricus rubellus* and *Dendrodrilus rubidus* from contaminated mine wastes. *Environmental Toxicology and Chemistry*, Vol 20, pp: 2336–2341, ISSN 15528618

Lukkari ,T., Taavitsainen, M, Väisänen, A. & Haimi, J. (2004). Effects of heavy metals on earthworms along contamination gradients in organic rich soils. *Ecotoxicology and Environmental Safety*, Vol. 59, No. 3, (November 2004), pp. 340-348, ISSN 01476513

Ma, W-C. (1982). The influence of soil properties and worm-related factors on the concentration of heavy metals in earthworms. *Pedobiology*, Vol 24, pp:109–119, ISSN 0046225X

Ma, W-C. (2004). Estimating heavy metal accumulation in oligochaete earthworms: a meta-analysis of field data. *Bulletin of Environmental Contamination and Toxicology*, Vol 72, pp: 663–670, ISSN 14320800

Manly, R. (1996). Biological indicators, In: *Environmental Analytical Chemistry*, F.W Fifield & P.J Haines, pp. (249 -274), Blackie Academic & Professional, ISBN 0632053836, London

MBS (Malawi Bureau of Standards). (2005). *Drinking Water Specification,* MS 214:2005, First Revision

McCornick, P.V. & Cairns, J. (1994). Algae as indicators of environmental change. *Journal of Applied Phycology*, Vol. 6, No. 5-6, pp.509-526, ISSN 09218971

Morgan, J.E. & Morgan, A.J. (1988). Earthworms as biological monitors of cadmium, copper, leadand zinc inmetalliferous soils. *Environmental Pollution*, Vol. 54, No. 2, (March 1988),pp. 123-138, ISSN 02697491

Neuhauser E.F., Cukic Z.V., Malecki M.R., Loehr R.C. and Durkin P.R., (1995). Bioconcentration and biokinetics of heavy metals in the earthworm. *Environmental Pollution*, Vol. 89, No. 3, pp. 293-301, ISSN 02697491

NSW (New South Wales). (2002). Blue green algal bloom management, In: *New South Wales Regional Algal Coordinating Committee*, Accesed on 21/06/2011,Availablefrom: <http://www.murraybluegreenalgae.com/whattodonext.htm#STEP%201>

OIS. (1998). Swiss Ordinance relating to Impacts on the Soil: 1st July 1998. SR 814.12 Switzerland, 1998.

Panda, R., Pati, S.S., Sahu, S.K. (1999). Accumulation of zinc and its effects on the growth, reproduction and life cycle of *Drawida willsi* (Oligochaeta), a dominant earthworm in Indian crop fields. *Biology and Fertility of Soils*, Vol 29, pp: 419-423, ISSN 01782762

Peijnenburg, W.J.G.M., Posthuma, L., Zweers, P.G.P.C., Baerselman, R., de Groot, A.C., Van Veen, R.P.C., Jager, D.T. (1999a). Relating environmental availability to environmetal bioavailability: soil type dependent metal accumulation in the oligochaete *Eiseniu undrei*. *Ecotoxicology and Environmental Safety*, Vol 44, pp: 294-310, ISSN 01476513

Peijnenburg, W.J.G.M., Posthuma, L., Zweers, P.G.P.C., Baerselman, R., de Groot, A.C., Van Veen, R.P.C., Jager, D.T. (1999b). Prediction of metal bioavailability in Dutch field soils for the oligochaete *Enchytraeus crypticus*. *Ecotoxicology and Environmental Safety*, Vol 43, pp· 170-186, ISSN 01476513

Rajfur, M., Klos, A, Waclawek, M. (2010). Sorption properties of algae *Sprogura* sp. And their use for determination of heavy metal ions concentrations in surface water. *Biolectrochemsitry*, Vol 80, pp:81-86, ISSN 15675394

Ramelow, G.J., Maples, R.S., Thompson, R.L., Mueller, C.S., Webre, C. & Beck, J.N. (1987). Periphyton as monitors for heavy metal pollution in the Calcasieu River estuary. *Environmental Pollution*, Vol. 43, No. 4, pp. 247-261, ISSN 02697491

Spurgeon, D.J., Hopkin, S.P. (1999). Tolerance of zinc in populations of the earthworm *Lumbricus rubellus* from uncontaminated and metal-contaminated ecosystems. *Archives of Environmental Contamination and Toxicology*, Vol 37, pp: 332–337, ISSN 14320703

Spurgeon, D.J., Svendsen, C., Rimmer, V.R., Hopkin, S.P., Weeks, J.M. (2000). Relative sensitivity of life-cycle and biomarker responses in four earthworm species exposed to zinc. *Environmental Toxicology and Chemistry*, Vol 19, pp: 1800–1808, ISSN 15528618

Sajidu, S.M.I., Masamba, W.R.L., Henry, E.M.T. & Kuyeli S.M. (2007).. Water quality assessment in streams and wastewater treatment plants of Blantyre, Malawi. *Physics and Chemistry of the Earth*; Vol 32, No. 15, pp. 1391–1398, ISSN 14747065

United States Environmental Protection Agency (US-EPA). (2010). Target analyte metals (Heavy metals) and cyanide, Accessed on 11/01/10, Available from; http://www.epa.gov/reg3hwmd/bf-lr/regional/analytical/metals.htm Van Vliet, P.C.J., Didden, W.A.M., Van der Zee, S.E.A.T.M., Peijnenburg, W.J.G.M. (2006). Accumulation of heavy metals by enchytraeids and earthworms in a floodplain. *European Journal of Soil Biology*, Vol 42, pp: S117–S126, ISSN 11645563

Van Vliet, P.C.J., van der Zee, S.E.A.T.M., Ma, W-C. (2005). Heavy metal concentrations in soil and earthworms in a floodplain grassland. *Environmental Pollution*, Vol 138, pp: 505–516, ISSN 02697491

Vijver, M.G., Vink, J.P.M., Miermans, C.J.H., van Gestel, C.A.M. (2007). Metal accumulation in earthworms inhabiting floodplain soils. *Environmental Pollution*, Vol 148, pp: 132–140, ISSN 02697491

Vyas, N.B., Spann, J.W., Heinz, G.H., Beyer, W.N., Jaquette, J.A., Mengelkoch, J.M. (2000). Lead poisoning of passerines at a trap and skeet range. *Environmental Pollution*, Vol 107, pp: 159–166, ISSN 02697491

Walkley, A.J., Black, I.A. (1934). An examination of the Degtjareff method for determining soil organic matter and a proposed modification of the chromic acid titration method. *Soil Science*, Vol 37, pp: 29-38, ISSN 15389243

Weltje, L. (1998). Mixture toxicity and tissue interactions of Cd, Cu, Pb and Zn in earthworms (Oligochaeta) in laboratory and field soils: a critical evaluation of data. *Chemosphere*, Vol 36, pp: 2643–2660, ISSN 00456535

WHO (World Health Organization), (2006). Guidelines for drinking water quality 3rd Edition, Accessed on 11/06/2010,Available from: <http://www.who.int/watersanitationhealth/dwq/gdwq3rev/en/ index html>

Wright, M.A. & Stringer, A. (1980). Lead, zinc and cadmium content of earthworms from pasture in the vicinity of an industrial smelting complex. *Environmental Pollution Series A, Ecological and Biological* ,Vol. 23, No. 4, (December 1980), pp. 313-321, ISSN 01431471

Żbikowski, R., Szefer, P., Latala, A. (2007). Comparison of green algae *Cladophora* sp. And *Enteromorpha* sp. as potential biomonitors of chemical elements in the southern Baltic. *Science of the Total Environment*, Vol 387, pp: 320-332, ISSN 00489697

2

Interaction Processes Between Key Actors – Understanding Implementation Processes of Legislation for Water Pollution Control, the Israeli Case

Sharon Hophmayer-Tokich
The Twente Centre for Studies in Technology and Sustainable Development (CSTM)
University of Twente
The Netherlands

1. Introduction

Israel is a semi-arid water-scarce country with currently ca. 250 cubic meters per capita per annum. With limited natural water sources due to the country's climate, geography and hydrology, it now faces its worst water crisis following several multi-year cycles of droughts (Israeli Ministry of Environmental Protection, n.d.). While these consecutive droughts aggravated the water-scarcity situation, it is by far not the main factor in the current crisis. In 2001 a parliamentary enquiry committee was established to investigate the water crisis. In its report the committee concludes that the crisis is mainly the result of inadequate management of the water sources throughout the years rather than the country's natural conditions. Focused on rapid development, the water sources were managed mainly for quantity and over-exploitation was the main answer to the growing water demand, depleting the country's water sources (Israeli Parliament, 2002). This created a major water deficit currently in an amount that is equal to the annual consumption of the country (Israeli Ministry of Environmental Protection, n.d.). The Committee's conclusions are not a new revelation. In fact this was reported in numerous experts' and State Comptroller reports in the past, though to no avail (Israeli Parliament, 2002; Adam, 2000).

In light of the natural water scarcity and the ever-growing demand for water, one would have expected that the quality of the water sources is well protected. However, in addition to the long-term over-exploitation, on-going uncontrolled pollution further deteriorated the water sources and their quality (Adam, 2000; Laster & Livney, 2009), with the main source of pollution being untreated or partially treated wastewater.

Whereas most of the population has received adequate sewerage facilities – removing wastewater from the population centers from early stages, wastewater treatment facilities lagged behind. In 1971 only 37% of the generated wastewater was treated, mostly by primary treatment, and in 1982 only 55%. By the end of the 1980s over 20% of the wastewater generated was discharged untreated into the environment, mainly to the adjacent dry river-beds, whereas the rest was mostly insufficiently treated resulting in low-quality effluent (The State Comptroller, 1991). The country's streams became in fact conduits of wastewater (Gasith & Pargament, 1998), contributing to pollution of the surface and groundwater resources.

Prevention of water pollution by wastewater receives, however, much attention during the 1990s, and a new trend can be seen in which most municipalities begin engaging in building advanced wastewater treatment plants (Gabbay, 2002). By 2008, for example, ca. 92% of the generated wastewater was treated, of which 55% to secondary level and additional 32% to tertiary level. While further improvement is yet to be made with the remaining 8% being untreated (Israeli Ministry of Environmental Protection, n.d.), this is, no doubt, a substantial improvement in comparison to previous decades.

The on-going water pollution was not the result of lack of legal tools. On the contrary, Israel has extensive legislations sufficient to protect its water sources and prevent their pollution. These were put in place at early stages of the State. It was their lack of enforcement that resulted in continuous pollution (Adam, 2000). Accordingly, while several factors can explain the shift observed starting the 1990s, forceful enforcement of the existing laws for the first time, is a crucial one (Hophmayer-Tokich, 2010). The question arises, why in previous decades were the laws not enforced, whereas starting the 1990s, they were? Laster (1976), addressed this issue in his doctoral dissertation, but at that point in time could refer only to the phase of lack of implementation. He referred to the institutional structure and particularly to the Ministry of Agriculture's vast authority over water management. Adam (2000) concludes that in addition to the governmental institutional failure, public indifference to the on-going pollution, also explains well the lack of implementation.

This chapter builds up on these previous works to further answer the question from actor-centered approach. A study of the historical development of wastewater resource regime in Israel reveals that, among other things, the relevant actors in the policy network and the power-imbalance between them explain well the long-term neglect as well as the paradigm shift (Hophmayer-Tokich, 2010). In this chapter this is further explored with the aim to analyze the interaction processes between the relevant actors in the policy network as explaining mechanism for implementation processes. This is done using the Contextual Interaction Theory. The analysis addresses implementation processes of water sources pollution control legislation, with focus on domestic wastewater treatment – the main source of pollution. Section 2 presents the theoretical framework used for the purpose of this analysis; in Section 3 the relevant legislative framework for pollution control in Israel is presented and the interaction processes between key actors are analyzed; conclusions are drawn in section 4.

2. Theoretical framework and methodology

Laws and regulations are policy instruments, one of the traditional elements of public policy (Kissling- Näf & Kuks, 2004). As all other instruments, they are meant to reach policy goals (Linder & Peters, 1989). However, the actual outcome of a given policy does not always match the policy goals (Owens, 2008), as in practice implementation of policy instruments may be hindered or lacking. Thus, a distinction should be made between 'policy formation' and 'policy implementation' processes, and when looking at possible changes intended by a given policy, one needs to analyze the implementation of the policy instruments. This has, thus, been given a separate attention in policy studies. 'Implementation' in this context is seen as "processes that concern the application of relevant policy instruments" (Bressers, 2004: 284). Since implementation of policy instruments is usually the responsibility of relevant actors in the policy network, its processes can be seen as a social interaction between these key-actors (Bressers & Lulofs, 2010). This led to the development of the Contextual Interaction Theory.

The Contextual Interaction Theory (in its latest conceptualization/adaptations based on Bressers, 2004 and Bressers & Lulofs, 2010) focuses on policy implementation and perceives policy processes (including policy implementation) as actor-interaction processes, meaning, processes that are influenced by activities and interactions of the relevant actors. Actors are individuals, representing themselves or their organizations, and within the context of implementation process include the responsible government officials ("implementers") and the target group of the policy. The Theory's basic assumption is that the characteristics of the actors involved, particularly their motivation, information, and power – are crucial in understanding courses and outcomes of policy processes (Bressers, 2004). This is based on the acknowledgement that for the accomplishment of any given task one needs a motivating objective, expertise, and capacity/resources (Owens, 2008). Motivation can incorporate both internal/own goals (values, self-interests) as well as external factors (such as those from higher authorities). It can also be influenced by self-effectiveness assessment. According to this concept an actor can become de-motivated if he perceives his preferred course of action to be beyond his capacity (de Boer & Bressers, 2011). Information incorporates issues such as interpretation, frames of reference as well as knowledge and accessibility to information required for execution of the task. Power incorporates available resources and control/authority (Bressers & Lulofs, 2010). Owens (2008) analyzed implementation literature reflecting important implementation variables, and found that in dozens of them the important implementation variables can be directly linked to the characteristics of motivation, information and power thus validating these characteristics as suitable.

According to the Contextual Interaction Theory these characteristics and the interaction between them influence the standpoint of a given actor regarding the policy in question and in turn his position and activities within the interaction process with other actors in the policy network. The characteristics of the actors are also influenced by external contexts such as the specific context of the policy (former decisions, specific circumstances), the structural context of the governance regime, and the wider context (such as political, economic, cultural and others). The interaction between the key characteristics and between the actors in the policy process can also change over time (Bressers & Lulofs, 2010).

The theory further assumes that policy implementation includes not only achieving implementation but also avoiding implementation. Interaction types may include: **cooperation**, either active (when actors have joint ambition), passive (e.g. when one actor is impartial about this implementation) or forced (when passive cooperation is imposed by a forceful and dominant actor); **opposition**, when one actor attempts to prevent implementation by other actors; and **joint learning** when only insufficient information prevents implementation (Bressers, 2004; Owens, 2008). The theory also distinguishes between two situations: lack of (or insufficient) implementation and failed/inadequate implementation ('adequate' with respect to the specific policy goals) (Bressers, 2004). As such, the theory is suitable for actor-centered analysis such as the one carried out in this chapter.

To summarize, the characteristics of motivation, information and power of each actor and the dynamics between them influence the interaction process between the relevant actors (implementers and target group), which in turn influence the output and outcome of the policy process. Based on this, the Israeli case is analyzed. In this case the Theory is also used to analyze the dynamics between the key characteristics and actors as explaining changes in the process over time (e.g. the shift from lack of enforcement and continuous pollution to forceful enforcement and pollution control). The analysis includes the actors; the interactions between them; and the outcome as a result.

2.1 Methodology

The findings presented in this chapter are based on a doctoral research and thus form a part of a larger study. The data collection includes documents' review and analysis as well as in-debt semi-structured interviews. The former includes review and analysis of relevant documents such as correspondences between position holders within relevant Ministries and other actors, minutes of meetings, relevant legislation, etc., using the State Archive, relevant reports and literature. The latter includes interviews with relevant stakeholders such as Government officials from all Ministries and organizations involved, environmental Non-Governmental Organizations, experts from the academy, and private consultants, since the establishment of the State. This is found suitable for the qualitative approach used in the research. At points, the author relies on dated literature sources. This is due to the historical perspective and is used for the purpose of the analysis.

3. Water pollution control and its implementation, the Israeli case

In this section the relevant legislation is presented, following which the relevant actors and the interaction between them are analyzed as to explain the implementation processes of the pollution control legislation. It should be noted that the relevant legal system and policy-network are highly complex, and only the most relevant laws / regulations and actors related to domestic wastewater are presented and discussed.

3.1 Relevant legislative framework

As above-mentioned, Israel has extensive legislations that can ensure the protection of its water sources (Adam, 2000). In fact, according to Laster (2000: 437), "in its early years Israel promulgated some of the most forward looking legislation in the world concerning protection of water sources".

The two most relevant and direct laws regarding freshwater pollution prevention and (domestic) wastewater treatment are: the Water Law (1959) and the Local Authorities (Sewerage) Law (1962). Two additional laws that should be mentioned include the Public Health Ordinance (1940) and the Streams and Springs Authorities Law (1965).

The Water Law, promulgated in 1959, establishes the framework for the control and protection of Israel's water resources. It is *the* principle law regulating freshwater sources in Israel and is regarded by Laster (2000: 441) as "a brilliant legislative code to protect all aspects of Israel's water and the recipe for its proper management". The Law defines the water sources (natural or man-made, including wastewater), and their ownership (public property, subject to the control of the State) as well as creates Israel's water institutions. It creates the Water Commission (later on: Water Authority) and the position of the Water Commissioner (later on: the Director of the Water Authority) as the higher authority with respect to water management, giving him vast authority to manage the water affairs of the State (Laster, 2000; Laster & Livney, 2009). Significant sections of the Law deal with pollution prevention and control of all water sources. In 1971, the law was amended to include prohibitions against direct or indirect water pollution, regardless of the state of the water beforehand. This amendment empowers the responsible Minister (initially the Minister of Agriculture) to set water quality standards for all sources and to promulgate regulations to prevent water pollution (Gasith & Pargament, 1998; Katin, 1976). The Water Commissioner further receives sanctioning power over polluters. For example, the power to require any person polluting water source to repair the situation within a reasonable amount of time and at the expense of the polluter (art. 11). Specific to wastewater pollution, the Law authorizes the Water Commissioner to order any polluter to provide him with a disposal plan. Failure to submit a plan or deviate from the plan can result in fine and loss of water supply, except for drinking purposes. The Water Commissioner can also bring criminal charges against a polluter – e.g. a mayor (Laster & Livney, 2009; Katin, 1976).

The Local Authorities (Sewerage) Law was promulgated in 1962 to enable local municipalities to construct sewerage works. The Law prescribes the rights and duties of local authorities in the design, construction and maintenance of sewage systems (Israeli Ministry of Environmental Protection, n.d.). According to the Law a local authority may (and upon the demand of the Minister of Interior, must) install a sewerage system within its boundaries (Local Authorities (Sewerage) Law, 1962). The Law requires each local authority to maintain its sewage system in proper condition to the satisfaction of the health authority (Israeli Ministry of Environmental Protection, n.d.). Initially, the Law addressed the sewerage systems only; in 1972 it was amended to include also the construction of wastewater treatment plants.

The Public Health Ordinance (1940), based on its amendment of 1970, provides the Ministry of Health a framework for the protection of the quality of drinking water including at the water source. Later amendments further prohibit any activities that cause environmental nuisances, including pollution by sewage (the Public Health Ordinance, 1940). Several regulations under this Law were promulgated over the years including the 1981 regulations for 'effluent intended for use in irrigation', restricting irrigation with effluent in accordance with the treatment level and the type of crops. These regulations, however, did not specify the required effluent quality and were thus difficult to enforce. In 1992 regulations for effluent standards (base-line quality defining the permissible concentrations of organic matter and suspended solids) were promulgated by the Minister of Health. These are considered a milestone in wastewater treatment processes in Israel. In 2010 new regulations were promulgated by the Minister of Environmental Protection and the Minister of Health, to include stricter requirements for effluent quality. These regulations set much higher treatment levels in existing and future wastewater treatment plants than were previously in force, for unrestricted irrigation and discharge to rivers (Israeli Ministry of Environmental Protection, n.d.).

The Streams and Springs Authorities Law, established in 1965, empowers the responsible Minister (initially the Ministers of Agriculture and Interior – dual control) (Gasith & Pargament, 1998) to establish an authority for a particular stream or part of a stream, spring, or other water source (Israeli Ministry of Environmental Protection, n.d.). Once created, a stream authority has the power to abate sanitary hazards and prevent pollution of the stream.

3.2 Actors in the policy network

Several actors can be mentioned with respect to implementation of the relevant legislation: implementers of the policy, actors not directly participating in the process but facilitating and providing support to other actors, and the target group of the policy. Within the context of this analysis actors are meant as organizations or individuals representing their organizations.

The Water Commissionaire is the highest authority regarding management of the water sources including prevention of their pollution (until the transfer of the responsibility for prevention of water pollution to the Ministry of Environmental Protection upon its establishment in 1989). The Water Commissionaire, however, was – until 1996 – subordinated to the Ministry of Agriculture, appointed by and answers to the Minister of Agriculture. Furthermore, all but two of the Water Commissionaires since the establishment of the State were clear representatives of the agricultural sector. The Water Commissionaire of 1977-1981 was appointed as the Director General of the Ministry of Agriculture in 1980 and served as both positions for several months (Israeli Parliament, 2002). As such, and with regards to the analysis offered by this chapter regarding motivation, information and power of main actors, the Water Commissionaire and the Ministry of Agriculture are considered as one actor (until 1996).

The Ministry of Agriculture was initially entrusted with the ministerial responsibilities over the Water Law and the Water Authority/Water Commissioner. As such, the Ministry of Agriculture had the highest authority and influence over the protection of water sources. In addition, it had a shared responsibility over the Streams and Springs Authorities Law and some influence via the Local Authorities (Sewerage) Law, as plans for the establishment of wastewater treatment plants require the approval of the Minister of Agriculture (art. 13). It was also involved via the Public Health Ordinance: according to article 65 when establishing regulations regarding effluent use for irrigation or other economic activity the Ministry of Health is to consult with the Ministry of Agriculture. Therefore, this Ministry is one of the main implementers of the relevant legislations.

The Ministry of Health was entrusted with the responsibility to protect the county's drinking water and as such had various powers to control water pollution, mainly under Chapter 6 of the Public Health Ordinance (this was also transferred to the Ministry of Environmental Protection upon its establishment), defining sewage as "nuisance". It also has the responsibility for the quality and use of effluent for irrigation (Adam, 2000). Via the Local Authorities (Sewerage) Law it is involved in approving sewerage works. As such, was also in a position to prevent water pollution.

The Ministry of Interior, as the Ministry responsible for the municipal sector, was entrusted with the ministerial responsibility over the Local Authorities (Sewerage) Law and can demand that a local authority will construct a wastewater treatment plant (art. 2). In addition, together with the Ministry of Agriculture it initially had the responsibility over the Streams and Springs Authorities Law. The Ministry of Interior is also to be consulted with, according to the Public Health Ordinance, concerning issues with implications for the local authorities (art. 3a).

The Ministry of Environmental Protection was established in 1989. Upon its establishment it assumed the responsibility - previously under other ministries, mainly the Ministry of Agriculture and Ministry of Health, for protecting the water sources and preventing their pollution. It now had the ministerial responsibility over the water pollution sections of the Water Law (art. A1 "prevention of water pollution"); the sections related to nuisances in the Public Health Ordinance, some responsibilities regarding wastewater treatment via the Local Authorities (Sewerage) Law and over the Streams and Springs Authorities Law. As such it became one of the main implementers of water pollution control in the following years.

The Ministry of Infrastructures was established in 1996 and received the jurisdiction over the Water Authority/Water Commissioner and the Administration for the Development of Sewage Infrastructures. By its establishment the link between water management and the agricultural sector was broken for the first time.

The Treasury Ministry, while not being a formal implementer, had a facilitating role via budgets allocated for this purpose, as elaborated in the following.

The Prime Minister's Office during the Rabin Government (established in 1992) prioritized development of infrastructure, wastewater treatment included. This government promoted relevant organizational change, allocated high budgets, as well as provided financial incentives for municipalities to treat wastewater, as elaborated in the following. As such it had a strong supportive and facilitating role.

As for the target group, since the chapter focuses on pollution from untreated domestic wastewater, and as the legal responsibility for wastewater collection, treatment and sanitary disposal is of the local authorities, the target group is defined as the municipalities / mayors.

3.3 Interaction processes between key actors

With respect to pollution of water sources by wastewater, two phases can be defined based on the enforcement of the relevant laws: 1) 1948 (the establishment of the State) until 1989; and 2) 1989 until the present. In the following, each actor's motivation, information and power are analyzed to explain the interaction processes between the main implementers and the target group, which in turn can explain the implementation processes of the pollution control legislation.

3.3.1 1948 until 1989

This phase is characterized by lack of enforcement. Only five court cases were filed by the Ministry of Health during the 1960s of which three addressed pollution by domestic wastewater. These, as Adam (2000) notes, were all the work of one official and were a rare exception. The Water Law, with the vast authority it provides for pollution prevention, was never enforced and sanctions prescribed by it, such as bringing criminal charges against polluters, went unused (Laster & Livney, 2009). The Springs and Streams Authorities Law was not implemented until 1989 when a stream authority was created for the first time with the establishment of the Yarkon River Authority (Gasith & Pargament, 1998). Indeed, the grim condition of the water sources (especially the streams), spoke for itself. Since relevant legislation was in place, had there been enforcement, this would have not been the case.

The main actor with this respect is the Ministry of Agriculture, as established above. Entrusted with the responsibility for the Water Law and with the sanctioning authority that the Law provides, as well as via its roles in other relevant legislations, it had vast authority and power to implement and enforce pollution control. With the Water Commission subordinated to it, it also had access to the relevant expertise as well as the information regarding the state of the water resources. It is, after all, the Water Commissionaire that was empowered to manage the State's water resources and had the means to do so. But what about its motivation? In the new State, agriculture was a very important economic sector but more than this, it became a national and political objective behind which stood the Zionistic ideologies of settling the land, the right to work own land in own country, etc. As such, the Ministry of Agriculture enjoyed vast political support across the political spectrum. It is against this background, that management of the water sources was entrusted to this Ministry. The Ministry of Agriculture, in fact, became the most important actor with respect to water management (with responsibilities also over other important water legislations such as the Water Drilling Law, 1955, and the Water Metering Law, 1955) and according to the parliamentary enquiry committee, managed the water sources exclusively and almost nothing could have been changed in water management without its cooperation (Israeli Parliament, 2002).

This Ministry's prime interest and goal, however, is to promote the agricultural sector and production, not to preserve the water sources as such (Adam, 2000). In the semi-arid country, agriculture on large scale requires irrigation, and the agricultural sector is the largest water consumer with ca. 70% of the water allocation (Central Bureau of Statistics, 2010) and as such utilization of water sources in support of agricultural production, outweighed other water management aspects (Laster & Livney, 2009). With a strong agricultural lobby, the Water Commissionaires were mostly affiliated with the agricultural sector, as above-mentioned, with preference to short term agricultural interests over long term water considerations (Adam, 2000). As wastewater is defined by the Water Law as a water source it was indeed perceived in this period by the Water Commissionaires primarily as an additional (cheap/free-of-cost) water resource to be utilized by farmers. The position of the Ministry of Agriculture was that wastewater - being first and foremost a water source for irrigation, should be managed and fall within the authority of the Water Commission thus the Ministry of Agriculture (State Archive, container GL2033/22, 6.1.54). To reduce costs involved and encourage farmers to utilize effluent, the Water Commission's/ Ministry of Agriculture's officials advocated a method which became known as 'agro-sanitation'. According to this method low-cost and low-tech facilities for primary treatment such as oxidation ponds and often only reservoirs, were constructed, following which the low-quality effluent was to be utilized for irrigation with further natural treatment at the root of the plant and the soil (State Archive, containers G5117; GL2033/26; GL4647, various documents; Shtreit, personal communication, 24.7.01). This method, however, contributed heavily to the pollution of the water sources. Not only that it produced low quality effluent, but when not utilized for irrigation, surplus effluents were discharged into the nearest stream. The facilities themselves became a source of pollution as they were not properly maintained and upgraded. Furthermore, wastewater treatment for the purpose of pollution prevention was grossly neglected. Since budgets were limited, municipal plans for construction of treatment facilities that did not include concrete plans for effluent reuse for irrigation, were rejected and not funded despite continuous pollution to water sources or to the environment, for example, in the cases of the towns of Nahariya and Zichron Ya'aquov (Fleisher, personal communication, 15.5.01; Tal, personal communication, 14.5.01; State Archive, container GL2102 4/24(6), 11.4.67). All these resulted in steadily growing pollution. To conclude, the prime interest of the Ministry of Agriculture – the most influential actor, conflicted with the interests of preserving the water sources, and within this inner conflict of interest, the prime interests related to utilization of wastewater, prevailed. As such, this Ministry had vast power and information but no motivation to enforce pollution control laws.

The Ministry of Health was another important actor via the sanitary aspect of wastewater treatment and its responsibility to protect drinking water and public health. The Ministry, however, had insufficient information, power (in practice) and motivation to enforce the laws. Regarding information, the Local Authorities (Sewerage) Law, for example, requires that a municipality maintains its sewage system in proper condition 'to the satisfaction of the health authority'. The Law, however, does not specify the type and amount of compliance required, nor does it specify which sanctions can be imposed, making its enforcement difficult. The Ministry of Health lacked also sufficient power to enforce the laws. Documents and correspondents' analysis reveals an on-going dispute between the Ministry of Agriculture and the Ministry of Health with respect to the authority related to wastewater management. The Ministry of Health, however, was smaller and weaker in comparison to the Ministry of Agriculture and in practice had little influence over the matter (State Archive, container GL2033/22, various documents). For example, in 1953 the Minister of Agriculture established the Sewage Committee—an inter-ministerial committee that would coordinate the different positions regarding wastewater solutions, and approve sewage plans. However, the Committee was to report to the Minister of Agriculture and out of eleven governmental members of the Committee six represented the Ministry of Agriculture (including the Chair) and only two the Ministry of Health, reflecting the power imbalance. Eventually, despite the different positions, the Ministry of Health approved the low-tech low-costs solutions that were advocated by the Committee, revealing passive cooperation. With respect to financial resources, the Ministry of Health had no own budgets for this topic (State Archive, container GL7345 2/13, September 1970). Finally, the Ministry of Health also lacked sufficient motivation. The Ministry of Health had other priorities to look after and within the Ministry, wastewater treatment received low priority. Pollution prevention and wastewater treatment were not considered a prime objective of this Ministry (Fliesher, personal communication, 15.5.01; Marinov, personal communication 23.5.01; Shelef, personal communication, 15.7.01). For example, the Public Health Ordinance allows the Minister to promulgate relevant regulations. However, the relevant regulations - 'effluent intended for use in irrigation', were only promulgated in 1981. The regulations restrict effluent irrigation but with respect to protection of public health, not the water sources, and were promulgated only in relation to a cholera outbreak that occurred in 1970 due to consumption of raw vegetables that were irrigated with untreated wastewater. The regulations allow irrigation with effluent based on permit system and for crops that are not meant for human consumption only, thus allowing irrigation of other crops. Moreover, the regulations did not specify the required effluent quality and merely prescribe that the dissolved oxygen concentration will be at least half a milligram per liter and that the effluent must not contain toxic compounds that may danger, in the view of the Director General of the Ministry of Health, the health of those that who come in contact with the effluent or with the irrigated crop (Gasith & Pargament, 1998). The fact that such regulations were promulgated at a relative late stage, do not address pollution of water sources and do not specify the required effluent quality, all reflect the lack of interest thus motivation to enforce water pollution control, by this Ministry. This lack of motivation can further explain the Ministry's weaker position in the actor's network and thus its passive cooperation.

The Ministry of Interior has a central role in the enforcement of the Local Authorities (Sewerage) Law and as such could have been an important actor. Entrusted with the authority to order municipalities to install proper treatment facilities, this Ministry had the power to enforce the Law and protect the water sources from pollution. The Ministry of Interior lacked, however, sufficient information. While the Ministry was represented in inter-ministerial committees such as the Sewage Committee, in practice it was a marginal actor. It lacked the technical and professional expertise and accepted the position of the Water Commissionaire and the Ministry of Agriculture with respect to their approach to wastewater treatment (Hecht, personal communication, 12.6.01). For example, following the cholera outbreak the government established a new inter-ministerial committee for wastewater management which was meant to define relevant policy, as well as the National Sewage Project – its operative arm, meant to execute this policy. The Director General of the Ministry of Interior was appointed as the formal Chair of the inter-ministerial committee based on the Ministry's responsibility for the municipal sector and as wastewater treatment is an obligation of the municipalities. This, however, was a formal appointment only; the Director General admitted to have little information on the subject and in practice it was the Water Commissionaire that ran the committee (Kantor, personal communication, 8.5.01). It was the Water Commissionaire that also ran the operative arm and controlled the budgets. Most importantly, though, the Ministry of Interior lacked motivation. According to Adam (2000), the use of the authority given to it by the Local Authorities (Sewerage) Law would have prevented or reduced water pollution caused by discharge of untreated wastewater to the streams. However, she notes, the Ministry of Interior chose not to exercise this authority. Being the Ministry responsible for the municipal sector, the Ministry's primal goal is to support the local authorities. The low-tech low-cost facilities that were advocated by the Ministry of Agriculture meant lower costs for the municipalities and with little interest and expertise in this topic, the Ministry of Interior accepted them as suitable solutions (Hecht, personal communication, 12.6.01). An example of the Ministry's prime interest in the state of the municipalities rather than in wastewater treatment can be seen in the following. Following the cholera outbreak, the Israeli Government signed in 1972 an agreement with the World Bank concerning a loan for the purpose of upgrading wastewater treatment facilities. As per this agreement, the World Bank required, among other things, that the local authorities will ensure that the funds they raise with wastewater charges will be used for wastewater management, allowing them to repay the loan as well as maintain and operate the facilities. To comply, the Ministry of Interior issued an order by the General Director stating that local authorities are required to do so. This, however, was never enforced (Shtreit, personal communication, 24.7.01). In practice many of the local authorities used income from water and wastewater charges for other purposes, especially when in fiscal stress (Laster & Livney, 2009). Both the Ministry of Interior and the Treasury Ministry realized that enforcing this order would mean financial burden for the local authorities, and ignored it (Hecht, personal communication, 12.6.01). This, too, reveals that the Ministry of Interior's prime interest was to support the local authorities and not to protect the water sources.

An additional actor that should be mentioned is the Environmental Protection Service. The Service was established in 1973 as a department in the Prime Minister's office following the Stockholm Declaration of 1972. In 1976 it was transferred to the Ministry of Interior, mainly in order to affect local authorities' handling of sewage. The Environmental Protection Service, however, had mainly a research position and while it was represented in scientific forums, it had no decision-making power (Marinov, personal communication 23.5.01). As such it had the motivation and the information, but not the power. Nonetheless, it prepared the ground for the Ministry of the Environmental Protection that would be established at a much later stage.

The Treasury Ministry has a facilitating role via budgets. However, prior to the agreement with the World Bank, separate national budgets were not allocated for wastewater treatment. The position of the Treasury Ministry was that the local authorities are expected to finance sewerage works by themselves, and that the wastewater solutions should be the cheapest ones available (State Archive, container GL4647/746, 11.12.1961). Following the cholera outbreak and per the agreement with the World Bank, the Treasury Ministry was to provide 60% of the projects' costs. With a clear interest to reduce costs, it followed the approach for low-cost facilities led by the Ministry of Agriculture.

The target group is defined as the local authorities and their mayors, as they were given the formal responsibility for wastewater treatment and could be legally charged for pollution of the water sources. The municipalities had insufficient information to carry out this task. The Local Authorities (Sewerage) Law, as established above, did not specify the type and amount of compliance required, as long the health authorities were 'satisfied'. In addition, most of the local authorities in these years were small and had little or no access to the required technical expertise (own personnel), and relied heavily on relevant central authorities such as the Sewage Committee. More importantly, they had insufficient resources. Prior to 1972 the government did not allocate separate budgets for wastewater treatment. Such funds became available after 1972 following the agreement with the World Bank, but these were assigned to the Central Government (the National Sewage Project, chaired by the Water Commissioner), which decided how to allocate the funds. Municipalities remained dependent on the central authorities to access these public funds. Raising private capital required the approval of the Treasury Ministry. This, however, was not considered appropriate during this phase. The only mayor at the time, trying to raise private capital, was the mayor of the coastal city of Haifa. Initiating the construction of a biological treatment plant during the 1950s, this mayor was a front-runner and ahead of his time. After establishing an inter-municipal cooperation and preparing the technical plans for the treatment plant, the final hurdle to overcome was the needed funds. The mayor, being extremely committed, managed to interest two French companies that would construct the plant with their own capital. In fact, a pioneer Build Operate Transfer (BOT) construction. The Treasury Minister, however, refused to approve this agreement, and only due to the mayor's determination, this was eventually approved. The Haifa treatment plant went into operation in 1961, decades before other municipalities followed. Other less determined and less capable mayors – would have not succeeded (Hophmayer-Tokich, 2005). And finally, the local authorities had no motivation. Most of the municipalities had little or no inner motivation and focused on housing and economic development. Sewerage systems were constructed in early stages to remove hazards from the population centers thus any pollution that was caused by wastewater, was – in most cases, not felt by the inhabitants. In the cases that the inhabitants did experience the nuisances, such as the city of Tel Aviv via the Yarkon Stream or the city of Hadera via the Hadera Stream, mayors were more pressed to take action. External pressure, by higher authorities, e.g. by law enforcement – was completely non-existent. The central authorities reflected that the lack of action was tolerated. Therefore, the target group had little information, very few capacities (power) and hardly any motivation to comply with the Law and treat wastewater. The main findings for this phase are presented in table 1.

To conclude, several actors are given the authority to implement water pollution prevention, but none of them do so. Some actors lack access to power and information while others have, however none of them have the motivation. The main reason for that is that protection of water sources is not the primal goal of any of them. The Water Commissionaire at this phase has the formal task and authority to protect the water sources and prevent their pollution, thus given the power and the information. However, subordinated to the Ministry

Actors	Motivation	Information	Power	Interaction processes
Ministry of Agriculture	Lacking; Primal goal – support agricultural production, not preserving the water sources. As such, advocated utilization of low quality (free of charge) effluent in support of agricultural production; No motivation to enforce wastewater treatment.	Via the Water Commission subordinated to it - has the expertise as well as the information regarding the state of the water resources	Authority/sanctioning (enforcement) resources as well as sufficient financial and personnel resources; Level of general political support – very high	Most powerful actor; However its power allows this Ministry to choose not to implement the Water Law or to implement according to own goals. This actor dominated the policy network based on its own goals and the wastewater management approach it advocated
Ministry of Health	Lacking; Pollution prevention and wastewater treatment – not primal objective, receives low priority within the Ministry	Lacking; the Sewerage Law does not specify the type and amount of compliance required, making its enforcement difficult	Lacking; insufficient enforcement resources; authority in practice; and financial resources aimed at wastewater treatment	Weak position in the actors network; little influence over the process due to passive cooperation
Ministry of Interior	Lacking; Primal goal – support the local authorities. Enforcing the Laws would have caused fiscal stress to the municipalities, which the Ministry of Interior was trying to avoid. Thus, no motivation to enforce wastewater treatment	Lacking: no expertise and technical know-how	Had the power via the Local Authorities (Sewerage) Law	Impartial actor (by choice); little influence over the process
Environmental Protection Service	High interest in protecting water sources	Has access to sufficient information via its experts and scientists	Has no power over the decision making or implementation processes	Marginal actor in the policy network
Target group: Municipalities / mayors	Low motivation to treat wastewater: little inner motivation, no external pressures, low self-effectiveness assessment	No specification of type and amount of compliance required (by the Sewerage Law) limited personnel	Very limited financial resources (no direct access to public nor private capital)	Weak position in the actors network; little influence over the process

Table 1. Characteristics and interaction processes, key actors; 1948- 1989 (lack of implementation)

of Agriculture and appointed as a representative of the agricultural sector, lacks the motivation to enforce the Law. The Ministry of Agriculture's prime values and interests are to promote agriculture, and securing water for irrigation is a high priority. Implementation of pollution control instruments is not perceived as contributing to these goals. This Ministry is the most powerful actor in the policy network and in fact used its access to power and information to dominant the process for its own interests. Within the interaction process, this Ministry was not keen on sharing the vast powers it possessed over the management of the water sources. According to Laster and Livney (2009), for example, Stream Authorities were not established as the Water Commissioner did not intend to share his power with another authority and the Minister of Agriculture supported this position. All this reflect the influence of this Ministry over the implementation process. Other actors with some responsibilities and authority could have improved the state of affairs, but would or could not do so. They lack the motivation, information, and power in various degrees and with the Ministry of Agriculture being the most powerful actor, consciously allow it to take the lead as wastewater treatment is not a primal goal of any of them. Most of them, especially the Ministry of Interior and the Treasury Ministry, accept the approached advocated by the Ministry of Agriculture as it suits their own interests, and the only actor that shows some attempts to object – the Ministry of Health, has a weaker position in the interaction process and eventually passively cooperates with the Ministry of Agricultures' approach. The target group lacks information and power, and with no access to funds was highly dependent on central authorities. Municipalities also lack motivation due to low internal motivation and lack of external pressure. Furthermore, mayors' lack of motivation can also be explained by the self-effectiveness assessment concept. With little influence over the process and its outcomes, even mayors that may have preferred to take action, such as in the cases of Nahariya and Zichron Ya'aquov, realized they have no capacity to take such actions and became de-motivated. The above-mentioned mayor of Haifa is a rare exception.

As a result, the implementation process can be defined as avoiding/lack of implementation due to passive cooperation. Most of the actors were impartial and passively cooperated with the standpoint of the more dominant actor – the Ministry of Agriculture, by not hindering nor stimulating the implementation of the Laws. As such, have little influence over the interaction process and thus on the policy outcome. The target group – is also in a weak position in the interaction between the actors and has little influence over the process as well.

3.3.2 1989 until the present

Starting 1989 a new trend of enforcement of relevant legislation can be seen. In response, most of the municipalities throughout the country, including ones that have neglected wastewater treatment for decades such as the city of Jerusalem, Be'er Sheva, Karmiel and others, began constructing highly advanced wastewater treatment plants. Several factors can explain this, of which the establishment of the Ministry of Environmental Protection is the most important one.

The Ministry of Environmental Protection was established in 1989 as a small Ministry, meant to provide a solution for a coalition crisis. It soon after, however, became a crucial actor in pollution control enforcement. As above mentioned, it assumed the responsibility for protecting the water sources and preventing their pollution from other Ministries. As such, it has the power to enforce relevant laws. It also has the relevant information. Although the Ministry started as a small low-budgeted Ministry, it was staffed by highly trained team of professionals, also based on the Environmental Protection Service which prepared the ground and provided the newly established Ministry with the professional expertise needed for the new tasks. Therefore the Ministry had access to power and information. These, however, were proven in the previous phase to be insufficient in the absence of motivation. More importantly, thus, the establishment of this new Ministry created for the first time an institution with an exclusive mandate to and interest in protecting the environment and the water sources (Adam, 2000). Furthermore, the Ministry's personnel and especially its first General Director were fully committed to create a change, and the first Ministers were environmentally-oriented and supported the development of relevant strategies and policies (Adam, personal communication, 6.9.01). As such this Ministry had clear motivation to enforce the laws. Its immediate actions with this respect, clearly reveal that.

Acknowledging that previously enforcement was non-existent, the Ministry's first action was to begin developing a comprehensive policy of forceful enforcement and the enforcement mechanism. As municipal wastewater was the main source of pollution the Ministry turned to establish a policy against polluting local authorities. The main Law this policy could have been built on was the Water Law (art. A1). The maximum penalty for water pollution according to this Article, however, was very low: only 4,500 NIS*, with no provisions for imprisonment. These sanctions were insufficient to deter polluting local authorities. The Ministry's first action was to amend the Law to allow more meaningful penalties. The amendment was completed in 1991, following which any polluter could have been penalized with one year imprisonment or a fine of 150,000 NIS, and in case of continued violation of the Law - seven days of imprisonment and additional fine of 10,000 NIS for every day of continuous pollution after a written warning was issued†. Furthermore, the amendment to the Water Law included also a provision for citizen suits, allowing an additional route for enforcement. In parallel the Ministry's officials started issuing warning letters to mayors demanding that they take actions to prevent pollution (Adam, personal communication, 6.9.01; Adam, 2000). In many cases the warning letters were sufficient, but in others lawsuits were filed. By 1995 fifty one lawsuits were filed by the Ministry of Environmental Protection against polluters, of which seventeen were against local authorities for violations of the Water Law (State Comptroller, 1996). Additional enforcement measure that was taken by the Ministry of Environmental Protection was an administrative measure of refusal to permit housing of newly built housing units unless the local authority in question had a proper wastewater treatment plant or advanced plans to construct one. In the beginning of the 1990s massive waves of immigrants from the former Soviet Union immigrated to Israel resulting in rapid development in most of the local authorities. At this point in time, taking such administrative enforcement measures put heavy pressure on mayors to treat wastewater (Marinov, personal communication, 23.5.01). In addition, the Ministry promulgated regulations on a wide range of issues related to water pollution (Adam, 2000), further enabling enforcement. The Ministry also promoted a policy towards stream rehabilitation. This includes establishing local administrations for stream restoration (30 such administrations were established by 2008) and regulating effluent discharge to streams. The latter – non-existent prior to the establishment of this Ministry, aims to enable base-line flow when potable water is unavailable, by permits system and by requiring stricter effluent quality when discharged to streams (Israeli Ministry of Environmental Protection, n.d.).

---

* Approximately $ 450 at the time (Adam, 2000)

† In 2008 the Law was amended to further increase the fines to 350,000 and 23,000 for every day of continuous pollution.

The Ministry's policy of enforcement, however, was not easily implemented and met objections from other Ministries, such as the Ministry of Housing. Furthermore, until 1993 the Ministry of Environmental Protection did not have its own prosecutors and it relied on the Attorney General and the district attorneys to file lawsuits. These, however, were not keen on cooperating due to work load, low priority for environmental issues, and the availability of administrative measure of enforcement. Finally, bringing criminal charges against mayors can meet political objections and on occasions the Ministry's staff was unable to file suits against a mayor. Determined to enforce the laws and promote a change, the Ministry's officials turned to other solutions when needed, including transferring material to an environmental non-governmental organizatios so that it can file a civil law suit against the polluter, in case of political pressure against the Ministry of Environmental Protection taking such action. All these reveal that the newly established Ministry of Environmental Protection had a strong motivation to enforce water pollution control measures. This strong motivation enabled the Ministry's officials to strengthen the power and information initially available to them by amendment of the Water Law to create a meaningful enforcement mechanism. Small and low-budgeted, and often facing political opponents and objections by other Ministries, the Ministry's strong motivation to enforce the Law gave it a strong position in the policy network.

The Ministry of Health became a more meaningful actor in this phase. In 1992 the Ministry promulgated the base-line quality regulations for effluent standards. These regulations served as an important tool for enforcement as for the first time the local authorities could have been required to treat wastewater to meet clear standards. The new regulations compelled municipalities to establish advanced wastewater treatment plants in order to meet these standards. As such these regulations provided the Ministry of Health more access to power – with more authority at hand, and information with a clear frame of reference for enforcement. It should be noted that the regulations were co-initiated by the Ministry of Agriculture which by the end of the 1980s acknowledged for the first time the need to divert high quality effluent for non-restricted irrigation for the agricultural sector (Hophmayer-Tokich, 2010). In 2010 the regulations were amended by the Ministry of Health and the Ministry of Environmental Protection to include stricter standards. With respect to motivation, it seems that in this phase, the Ministry of Health has higher motivation to enforce the law. This can be explained by the drastic reduction in the power of the Ministry of Agriculture on one hand, and with the establishment of the Ministry of Environmental Protection on the other. By the end of the 1980s, in the fast growing market economy, agriculture lost both its ideological status and its economic significance. As such, the agricultural sector and the Ministry of Agriculture lost its political power. Regarding water management, with the transfer of the responsibility for water pollution control to the Ministry of Environmental Protection, and in 1996 with the transfer of the Water Commission (by then: Water Authority) to the newly established Ministry of Infrastructure, the Ministry of Agriculture lost most of its power over water management. This cleared the way for the Ministry of Health – traditionally a weaker opponent and passive cooperator of the Ministry of Agriculture with respect to wastewater treatment, to be in a stronger position to enforce the relevant laws. On the other hand, the establishment of the Ministry of Environmental Protection and the transfer of some of the Ministry of Health's authority over water pollution control and wastewater treatment to this Ministry, resulted in a new rivalry, this time between these two Ministries. It seems that this rivalry and the enforcement actions taken by the Ministry of Environmental Protection gave the Ministry of Health additional motivation to enforce the laws and establish its position with this respect, as well (Tal, personal communication, 14.5.01; Balasha, personal communication, 14.5.01). As such, the Ministry of Health in this phase has better access to information and power, and higher motivation to enforce wastewater treatment.

The Ministry of Agriculture, as abovementioned, lost most of its power, both in general and specific to water pollution control. In addition to the above-mentioned, in 1992 under the Rabin Government the Administration for the Development of Sewage Infrastructure was established to replace the previous inter-municipal Sewage Committee and its operative arm. In contrast to the past, whereby the Water Commissionaires were traditionally – officially or not - chairing the National Sewage Committee, the new Administration was subordinated to the Ministry of Interior and represented by different Ministries, excluding the Ministry of Agriculture (State Comptroller, 1996). Furthermore, its operative arm was headed by professional wastewater engineers with a clear affiliation to advanced technologies rather than to agricultural interests. The advocated technical solutions were now advanced treatment technologies (Gurion, personal communication, 16.07.01). All these resulted in loss of power by this Ministry in this phase. At the same time, an interesting development is that the Ministry of Agriculture – for the first time, shows more motivation for a higher effluent quality and high level treatment, from its own interests. By the end of the 1980s consecutive droughts resulted in major cut-downs in potable water allocation for irrigation. At the same time, the global price of cotton – previously a very lucrative crop, dropped drastically. This forced farmers to shift to more profitable crops, which require high quality effluent. These two resulted in high interest of the agricultural sector in a reliable alternative source of water for non-restricted irrigation – high quality effluent. This explains the co-initiation of the base-line effluent quality regulations. To conclude, at the start of this phase a shift in the Ministry's motivation is observed, revealing higher motivation to treat wastewater to a high level, in combination with drastic reduction in power, thus much less influence over the enforcement process and lower position in the actors' network.

The Prime Minister's Office should be mentioned with respect to Prime Minister Rabin. The Rabin Government, established in 1992, prioritized development of infrastructure in general, wastewater infrastructure included, in light of the massive waves of immigrants from the former Soviet Union. Apart from establishing the Sewage Administration, mentioned above, this government allocated substantially increased budgets for wastewater treatment plants. If prior to 1992 the annual budget for wastewater treatment was 15-20 million NIS, this grew to 180 million NIS in 1993, 250 million NIS in 1994, 450 in 1995 etc. Moreover, in order to provide incentives for municipalities to engage in advanced wastewater treatment and apply for loans, partial grants up to 25%[‡] of the overall loan became available for municipalities that submitted plans and received their approval within the first three years (Reich, personal communication, 14.06.01). This provided local authorities with higher motivation on one hand, and sufficient resources (power) on the other.

With respect to the Ministry of Interior, its involvement in enforcement remains unchanged. The Ministry of Infrastructures was established in 1996 and received the jurisdiction over the Water Authority/Water Commissioner and the Administration for the Development of Sewage Infrastructures. This Ministry does not have power regarding enforcement of water

---

[‡] 25% for municipalities applying for a loan in the first year, 20% second year, 15% third year (Riech, personal communication, 14.06.01).

pollution prevention legislation, but has a clear mandate and objective to upgrade the State's infrastructure. Via the Sewage Administration it has control over the national budgets for wastewater treatment and has the professional staff to advice municipalities on that matter. As such it has a facilitating and supportive role.

The target group in this phase is facing a different situation. In terms of motivation, it now has a very strong motivation to treat wastewater, mainly due to external pressure from enforcement authorities. Many mayors received warning notifications, were warned that a law suite would be filed against them, and in some cases indeed faced lawsuits. In parallel, in the midst of a massive building and development phase, were not allowed to populate the newly built neighborhoods unless they had an existing or planned proper wastewater treatment plants. Mayors realized that for the first time not treating wastewater is no longer tolerated by the relevant authorities. At the same time municipalities are given the resources and the means to comply with the law. Regarding information, the municipalities now have clear standards to meet, thus have a frame of reference to what is asked of them. Most of the municipalities at this phase also have access to professional staff either via their own water departments, via inter-municipal cooperation, or private consultants. In any case, with the new Sewage Administration staffed with professionals not affiliated with the agricultural lobby, municipalities now have access to relevant information. And finally, municipalities now have access to financial resources, either in the form of loans from the government or by raising private capital which by the end of the 1990s was encouraged by the government and the Treasury Ministry. This gives municipalities more power in comparison to the previous phase, in which most of them were highly dependent on the central authorities. This can also be seen as influencing their self-effectiveness assessment, thus further motivating them.

To conclude, several factors co aligned to change the interaction process between actors and result in the paradigm shift: the establishment of the Ministry of Environmental Protection, the drastic reduction in the power and influence of the Ministry of Agriculture with – at the same time – its acknowledgement for the need to direct high quality effluent for irrigation, and the Rabin Government's policy of investment in infrastructure. All these shifted the balance and altered the interaction between the relevant actors and resulted in the shift. With the establishment of the Ministry of Environmental Protection finally there was an institution that has access not only to power and information but also to motivation. With the sole interest in protecting the water sources, and with highly motivated staff to create a change, this Ministry started a new trend that could have not been overturned. In parallel, the drastic reduction in the power of the Ministry of Agriculture, that previously controlled almost exclusively the water management sector, including wastewater treatment, paved the way to a new interaction between the actors. With the new interaction, it can also be seen that the establishment of a Ministry that was determined to make a change and enforce the laws, influenced other relevant actors, previously with a weaker or impartial positions, especially the Ministry of Health, to embark on this trend and enforce the laws. Be it due to the new rivalry over authority, or due to actual interest in the needed change, the Ministry of Health became a relevant actor in the new trend of enforcement. Finally, enforcing the laws without providing the means to comply with the law would have made it very difficult for the municipalities to establish advanced wastewater treatment plants. The Rabin's Government allocated high budgets and gave strong financial incentives for municipalities to embark on the new trend and invest in wastewater treatment; it also provided professional assistance and framework via the new Sewage Administration, now headed by wastewater engineers rather than representatives of the agricultural sector. This process is still on-going and with the stricter regulations recently promulgated by the Ministry of Environmental Protection and Ministry of Health, and with the further amendment of the Water Law in 2008 to increase the penalties that can be imposed on water polluters, it seems that the high motivation to improve the state of the water sources, continues.

As a result, in this phase the implementation process can be defined as achieving implementation. The interaction type, however, seems to not fully fit the types prescribed by the Theory. Since implementation is achieved, it can be described as cooperation as more actors share similar objectives. However, rivalry and power struggles hinder effective cooperation and coordination and it seems that the similar objectives are motivated by different reasons which result in more independent actions taken by different actors rather than active cooperation to achieve implementation. The Ministry of Environmental Protection's interest is to protect the water sources; the Ministry of Health aligns either due to its wish to establish its authority within the new rivalry with the Ministry of Environmental Protection over control or due to its own interest in protecting the public health; the Ministry of Agriculture's interest co-aligns as for the first time the agricultural sector acknowledges its need for high quality effluent for non-restricted irrigation as a reliable water source to replace potable water allocations for irrigation; the Prime Minister's Office prioritize infrastructure to support the absorbent of massive immigration wave. All these influence the interaction process and aligned to result in different actors taking actions, based on their different motivations and own-interest, to enforce the law and promote advanced wastewater treatment. Although the process was not always smooth and faced objections such as by the Housing Ministry and although the challenges in this field are far from being resolved, the new interaction process yielded enforcement trend and positive results. The target group was now being pressured into compliance and at the same time assumed the power needed for the task. With more access to information and resources, municipalities are now in a stronger position in the interaction process to influence outcomes. The main findings for this phase are presented in table 2.

Actors	Motivation	Information	Power	Interaction processes
Ministry of Environmental Protection	Clear ideology and unambiguous mandate to protect and preserve country's water resources. The first independent organization with the sole interest of protecting the environment	Highly trained and professional staff, experts and scientists, related to environmental as well as legal matters. The former Environmental Protection Service provided the needed base for the new Ministry	The Ministry assumes responsibilities and authorities previously belonging to other ministries. Despite little financial resources, the Ministry makes the most out of its authority/sanctioning/ enforcement powers; soon after its establishment amends the Water Law to provide a meaningful enforcement tool and take immediate enforcement actions	Became a crucial actor in the policy network in terms of law enforcement

Ministry of Health	Higher motivation in comparison to previous phase. Mostly, power struggle with the newly built Ministry of Environmental Protection as a motivation to become a more powerful actor	The base-line quality regulations, promulgated by the Ministry of Health in 1992, gave a frame of reference for law enforcement	Increased: Enforcement resources and authority in practice (via e.g. the base-line quality regulations); Still lacks own financial resources aimed at wastewater treatment	Stronger position in the actors' network especially with respect to law enforcement; more influence over the process; aligned with other actors to promote advance wastewater treatment
Ministry of Agriculture	For the first time acknowledges the need to utilize high quality effluent in support of agricultural production due to several drought years and cut-down in potable water allocation for agriculture, leading to co-initiation of the base-line quality regulations	With the government's decision from 1992 to establish the Sewage Administration, headed by professional wastewater engineers and with the transfer of the Water Authority to the newly established Ministry of Infrastructure in 1996 lost its grip over the relevant information	Level of general political support – very low; lost most of its powers in the new advance market-based economy. With the transfer of the responsibility for water pollution control to the Ministry of Environmental Protection in 1989 and the Water Authority to the Ministry of Infrastructure in 1996, lost most of the relevant authority and financial resources	Much reduced power; less influence over the process but aligned with other actors (from own goals) to promote advance wastewater treatment
Ministry of Infrastructure	Objectives: upgrade national infrastructures	Received the jurisdiction over the Water Authority and the Administration for the Development of Sewage Infrastructures and as such has access to relevant information	Received the jurisdiction over the Water Authority and the Administration for the Development of Sewage Infrastructures and as such has relevant financial and personnel resources	Interests aligned with other actors to promote advance wastewater treatment
Target group: Municipalities /mayors	High motivation to treat wastewater mainly due to external pressures (law enforcement)	clear specification of type and amount of compliance required (by the base-line quality regulations); better access to technical know-how	Better access to financial resources (both public and private)	Target group is pressured into compliance with the law on one hand and given access to resources for implementation on the other thus in a better position in the actors' network

Table 2. Characteristics and interaction processes, key actors; 1989-present (implementation)

4. Conclusions

In the State of Israel wastewater treatment has been neglected for decades. Municipalities – legally responsible for wastewater collection, treatment, and sanitary disposal, established the collection systems to remove hazards from their population centers, but in most cases neglected the elements of treatment and sanitary disposal. Untreated or partially treated wastewater was discharged into the environment, mostly to the nearest stream or dry riverbed, resulting in on-going pollution of the scarce water sources. Advanced legislation was in place since early stages of the State, providing the sufficient legislative framework to protect and preserve the water sources but the laws were not enforced. The intended policy goals were thus not attained due to lack of implementation. Since the beginning of the 1990s a shift can be seen and most municipalities, including ones that have neglected wastewater treatment for decades, began establishing advanced treatment plants. This is mainly the result of forceful enforcement and the pressure put on municipalities and their mayors. While water pollution by wastewater is far from being resolved, there is - no doubt, a substantial improvement in comparison to the past. This chapter aims to analyze and explain the lack of implementation of the policy instruments – relevant legislation for pollution control, as well as the shift to its implementation.

The implementation processes in the Israeli case are analyzed from actor-centered approach based on the Contextual Interaction Theory. This approach assumes that since implementation of policy instruments is the responsibility of relevant actors in the policy network, they can be seen as a social interaction between key-actors. The characteristics of the actors involved, particularly their motivation, information and power, are analyzed as to explain the standpoint and activities of each of the actors in the policy network. These further explain the interactions between the key actors and in turn, the outcome with respect to the implementation process. Accordingly, the analysis offered in this chapter addresses the key actors in the implementation processes of pollution control legislation, their key characteristics, and the interaction between them.

The Israeli case shows a shift from a phase of lack of implementation prior to 1989 to a phase of achieving implementation following 1989. It also reveals that the Contextual Interaction Theory can explain well these implementation processes. In the first phase relevant actors had access to information and power in various degrees but none of the actors with the power to enforce the law - had also the motivation to do so. The analysis shows that none of the actors has a primal interest in protecting the water sources by requiring municipalities to treat wastewater as all of them had other more important priorities to look after. As such, the most influential actor – the Ministry of Agriculture, dominated the policy network based on the lack of motivation and relative lack of power and information of other actors, and avoided implementation of water pollution control to serve its own primal interest – utilizing wastewater for agricultural production. This Ministry advocated a low-cost low-tech wastewater management approach that served the short term agricultural interests but not the long term water considerations. Other actors passively cooperated with this approach either because it served their own interests as well, e.g. in the case of the Ministry of Interior, or because they had a weaker position in the network to influence the outcome, e.g. the Ministry of Health and the target group. This influenced the outcome of the implementation process – insufficient wastewater treatment and on-going pollution. The interaction between the actors had changed starting 1989, most importantly due to the introduction of a new actor – the Ministry of Environmental Protection with a sole interest and high motivation to protect the water sources, as well as with a change in the key characteristics of the Ministry of Agriculture, with less power and higher motivation to promote advanced treatment. Wider contexts such as the massive immigration from the former Soviet Union and the rapid development it brought with it further influenced the implementation processes. In this new interaction, the Ministry of Environmental Protection promoted forceful enforcement and other actors participated in this process, e.g. the Ministry of Health, or supported it, e.g. the Prime Minister's Office, based on their own priorities. Putting pressure on the target group on one hand and providing it with means to comply on the other, gave municipalities motivation as well as information and power to comply. As a result, in this phase advanced wastewater treatment is achieved in many of the municipalities. Still facing many challenges, the enforcement trend continues.

Several lessons can be learnt from the Israeli case. (i) enforcement processes are indeed actor interaction processes and the result of this interaction in terms of the policy outcome can be well explained, thus predicted, by understating actors' access to motivation, information and power; (ii) key characteristics of the key actors can change over time and in turn result in changed outcome of the process; (iii) motivation seems to be the most important characteristic explaining implementation processes. Access to power and information alone were found to be insufficient as in the case of the Ministry of Agriculture. Furthermore, motivation can be used to create or increase access to better information and power as was the case with the Ministry of Environmental Protection. Understanding this, an intervention measure in cases of on-going pollution could be to create motivation either within the existing actors, e.g. the Ministry of Agriculture's acknowledgement of the need for utilization of high quality effluent for non-restricted irrigation, or by creating a new actor with high motivation, as was the case with the establishment of the Ministry of Environmental Protection; (iv) in the absence of an actor with access to both motivation and power, establishing an actor with motivation and information can in the long run prove to be useful as in the case of the Environmental Protection Service. While this actor had no power to enforce the law, it prepared the ground for such enforcement using its motivation and information and once an relevant organization with access to power was established – the Ministry of Environmental Protection, it could immediately embark on the tasks ahead using the information available by the Service; (v) once there is an actor with motivation and actions are taken, others may follow even if just to maintain and justify their authority over the matter. Pollution by untreated wastewater is a major environmental problem in many countries worldwide. Understanding these processes and interactions between key actors may assist predicting outcomes of policy processes and allow deliberate interventions to influence the motivations, information and power of the key actors.

5. Acknowledgements

The author is grateful to Nurit Kliot, the supervisor of the PhD study which the findings in this paper are based on.

6. References

Adam, R. (2000). Government Failure and Public Indifference: A Portrait of Water Pollution in Israel. *Colorado Journal of International Environmental Law and Policy*, 11. 2, pp. 257-376.

de Boer, C. & Bressers, H. (2011). Contextual Interaction Theory as a Conceptual Lens on Complex and Dynamic Implementation Processes. *Proceedings of Conference: Challenges of Making Public Administration and Complexity Theory Work*, Rotterdam the Netherlands, June 2011

Bressers, H. (2004). Implementing Sustainable Development: how to know what works, where, when and how, In: *Governance for Sustainable Development: The Challenge of Adapting Form to Function*, Lafferty, W.M. Ed., pp. 284-318, Edward Elgar Publishing , ISBN 1-84376-769-4, Cheltenham, UK & Northampton, MA, USA.

Bressers, H. & Lulofs, K. (2010). Analysis of boundary judgments in complex interaction processes, In: *Governance and Complexity in Water Management*, Bressers, H. & Lulofs, K. Eds., pp. 17-32, Edward Elgar Publishing , ISBN 978-1-84844-955-8, Cheltenham, UK & Northampton, MA, USA.

Central Bureau of Statistics. (2010). Water Production and Consumption table 21.5. *Statistical Abstract of Israel*, 61.   Jerusalem: Central Bureau of Statistics (in Hebrew) (information and data in English available from: http://www1.cbs.gov.il/reader/shnatonenew_site.htm)

Gabbay, S. (2002). *The Environment in Israel*; Ministry of the Environment: Jerusalem, Israel. Lase accessed on: 29 July 2011, Available from: http://www.sviva.gov.il/bin/en.jsp?enPage=e_BlankPage&enDisplay=view&enD ispWhat=Zone&enDispWho=Environment_Israel_2002&enZone=Environment_Isr ael_2002

Gasith, A. & Pargament, D. (1998). Practical obstacles to effective implementation of environmental enforcement: the case of the coastal streams of Israel, In: *Tel Aviv University Studies in Law*, 14. Pp. 117-134

Hophmayer-Tokich, S. (2005). The Construction of the Haifa Regional Sewage Treatment Plant – A Case Study. *Horizons in Geography*, 64-65, pp. 217-227 (in Hebrew)

Hophmayer-Tokich, S. (2010). The Evolution of National Wastewater Management Regimes - the Case of Israel. *Water*, 2. 3, pp. 439-460 doi:10.3390/w2030439

Israeli Ministry of Environmental Protection. (n.d). Official website. Lase accessed on: 29 July 2011, Available from: http://www.sviva.gov.il

Israeli Parliament. (2002). Report of the parliamentary enquiry committee for the water sector management (in Hebrew)

Katin, E. (1976). Environmental Law and Administration in Israel. *Environmental Policy and Law*, 2. pp. 94-97.

Kissling- Näf, I. & Kuks, S. (2004).Introduction to Institutional Resource Regimes: comparative framework and theoretical background, In: *The Evolution of National Water Regimes in Europe*, Kissling-Näf, I. & Kuks, S. Eds., pp. 1-23, Kluwer Academic Publishers, ISBN1-4020-2483-5, Dordrecht, the Netherlands.

Laster, R. (1976). *The Legal Framework for the Prevention and Control of Water Pollution in Israel*. Doctoral Dissertation, Hebrew University, Jerusalem

Laster, R. (2000). Catchment Basin Management of Water. *Water, Air, and Soil Pollution*, 123. pp 437-446

Laster, R. & Livney, D. (2009). Israel, the Evolution of Water Law and Policy, In: *The Evolution of the Law and Politics of Water*; Dellapenna, J.W. & Gupta, J. Eds., pp. 121-137, Springer, ISBN 978-1-4020-9866-6, Dordrecht, the Netherlands

Linder, S. H. & Peters, B. G. (1989). Instruments of Government: Perceptions and Contexts. *Journal of Public Policy*, 9.1, pp. 35-58.

Owens, K. (2008). *Understanding how actors influence policy implementation: A comparative study of wetland restorations in New Jersey, Oregon, The Netherlands and Finland.* Doctoral Dissertation, University of Twente, Enschede

The Local Authorities (Sewerage) Law. (1962). [unofficial English translation available at: http://www.sviva.gov.il/Enviroment/bin/en.jsp?enPage=e_BlankPage&enDispla y=view&enDispWhat=Object&enDispWho=Articals^l2410&enZone=wastew_law]

The Public Health Ordinance. (1940). (in Hebrew), available at: http://www.sviva.gov.il/bin/en.jsp?enPage=BlankPage&enDisplay=view&enDis pWhat=Object&enDispWho=Articals^l843&enZone=law_water

The State Comptroller. (1991). Establishment of sewage facilities. *Annual Report*, 41, pp. 424-433, ISSN 0334-9713 (in Hebrew)

The State Comptroller. (1996). Establishment of sewage facilities: collection, treatment and disposal. *Annual Report*, 46, pp. 518-537, ISSN 0334-9713 (in Hebrew)

The Water Law. (1959). [unofficial English translation available at: http://www.sviva.gov.il/Enviroment/bin/en.jsp?enPage=e_BlankPage&enDispla y=view&enDispWhat=Object&enDispWho=Articals^l2419&enZone=wat_law]

Archive sources:

State Archive, container G5117; various documents n.d.

State Archive, container GL2033/22, document dated 6.1.1954; various other documents n.d.

State Archive, container GL2033/26; various documents n.d.

State Archive, container GL2102 4/24(6), document dated 11.4.1967

State Archive, container GL4647, various documents n.d.

State Archive, container GL4647/746, document dated 11.12.1961

State Archive, container GL7345 2/13, document dated September 1970

Personal communication:

Adam, R. Legal Department, Ministry of Environmental Protection, personal communication via phone, 6.9.2001

Balasha, E. Wastewater Engineer, Balasha-Yalon consults, Haifa, Israel, personal communication, 14.5.2001

Fleisher, M. Former official in the Ministry of Health, Tel Aviv, Israel, personal communication, 15.5.2001

Gurion, Y. Former Chief Engineer, the Administration for the Development of Sewage Infrastructure, Haifa, Israel, personal communication, 16.07.2001

Hecht, A. Director of the Administration of Local Governance, the Ministry of Interior, 1965-1987, Tel Aviv, Israel, personal communication, 12.6.2001

Kantor, M. Water Commissioner 1959-1977, Ma'agan Michael, Israel, personal communication, 8.5.2001

Marinov, U. Former Director General, Ministry of Environmental Protection, Haifa, Israel, personal communication, 23.5.2001

Reich, B. Former Deputy General Manager, Administration for the Development of Sewage Infrastructures, Tel Aviv, Israel, personal communication, 14.06.2001

Shelef, G. Former official in the Ministry of Health, Haifa, Israel, personal communication, 15.7.2001

Shtreit, S. Director of the National Sewage Project 1972-1992, Tel Aviv, Israel, personal communication, 24.7.2001

Tal, M. Ministry of Health, District Engineer, Haifa District, Haifa, Israel, personal communication, 14.5.2001

3

Fluorescence Spectroscopy as a Potential Tool for *In-Situ* Monitoring of Dissolved Organic Matter in Surface Water Systems

Elfrida M. Carstea
*National Institute of Research and Development for Optoelectronics INOE 2000, Magurele, Ilfov
Romania*

1. Introduction

Water is a common substance, yet life cannot exist without it, being the major component of all living things. Considering the tremendous impact water has on life health, it is always an imperative task to study its quality. During the past decades, more advanced techniques were developed not only to generally characterise the water quality, but also to analyse DOM fractions.

Organic matter is present in every type of aquatic system and, due to the influence that it has on their ecological health, it can be used as a useful water quality indicator. The organic matter fraction from natural waters can be autochthonous, formed in situ through microbial activity, algal productivity, invertebrate grazing, etc., and allochthonous, formed externally and brought into the water system through soil leaching, geological activities or degradation of terrestrial vegetation (Winter et al., 2007). Human activities can influence both of these fractions: increased algal - derived organic matter due to eutrophication increased microbially - derived organic matter from human and animal wastes, and changes in allochthonous organic matter from changes in land use.

An emerging technique, fluorescence spectroscopy, which was successfully used in biology, medicine or chemistry, became a promising approach to the assessment of organic aquatic components and organic pollutants, due to its rapid analysis and high sensitivity. Fluorescence spectroscopy, in the form of three dimensional excitation-emission matrix (EEM), synchronous fluorescence spectrum (SFS) and laser induced fluorescence spectrum (LIFS) can be used to estimate water pollution and to probe the composition of DOM in watersheds. Although the fluorescence technique have been in the attention of those who are interested in real-time monitoring of water pollution, only few studies have been made in this field (Carstea et al., 2010; Downing et al., 2009; Spencer et al., 2007).

This paper proposes to review some of the methods potential to characterise different water systems that have dissimilar hydrological and geographical features and different sources of water pollution. Prior to this, theoretical aspects of fluorescence principles and dissolved organic matter properties will be shortly described.

2. Principles of fluorescence spectroscopy

Fluorescence is a special type of luminescence that describes the emission of light from molecules, named fluorophores, in electronically excited states. The fluorophores absorb energy in the form of light, at a specific wavelength, and release it in the form of emission of light, at a specific higher wavelength (i.e., with lower energy). The general principles of light absorption and emission can be illustrated by a Jablonski diagram, as seen in figure 1.

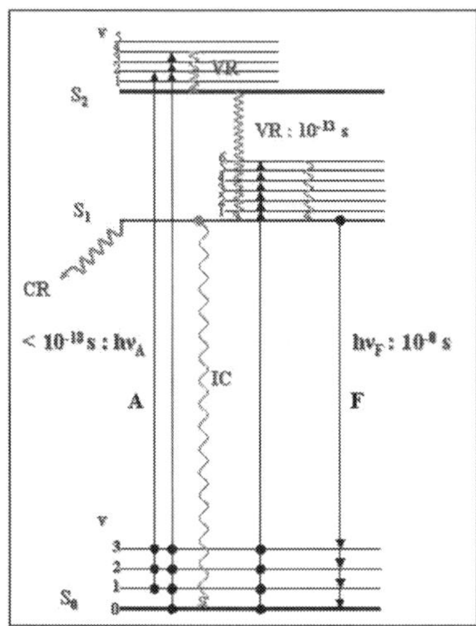

Fig. 1. Jablonski diagram presenting the processes of absorption (A) and fluorescence (F); VR – vibrational relaxation, IC – internal conversion.

When a molecule is found in the ground singlet state, S_0, and absorbs light, the light energy is transferred to the electronically excited states: singlet states, S_1 or S_2. Afterwards, the molecule is subjected to internal conversion or vibrational relaxation, which implies the transition from an upper electronically excited state to a lower one, lasting from 10^{-14} to 10^{-11} s. In the final stage, emission occurs when the molecule returns to the ground state, S_0, in 10^{-9} to 10^{-7} s, emitting light at a greater wavelength, according to the difference in energy between the two electronic states (Lakowitz, 2006; Valeur, 2001). This process is known as fluorescence. When excitation source is a laser, the fluorescence is called laser induced fluorescence.

The Jablonski diagram shows that the energy of the emission is generally less than that of absorption. Thus, fluorescence typically occurs at lower energies or longer wavelengths. This effect is called Stokes' shift, which is caused by several factors: the rapid decay to the lowest vibrational level of S_1, further decay of fluorophores to higher vibrational levels of S_0, solvent effects, excited-state reactions, complex formation, and/or energy transfer (Lakowicz, 2006).

Generally, the emission spectrum for a given fluorophore is a mirror image of the excitation spectrum. The symmetry is a result of the same transitions, which are involved in absorption and emission and the similarities of the vibrational levels of S_0 and S_1 (Christensen, 2005).

2.1 Factors affecting fluorescence intensity

The fluorescence excitation and emission spectra comply with the above-mentioned rule and properties, but several environmental factors can change the characteristics of the fluorescence signal. The fluorescence response is highly affected by solution temperature, composition, concentration, pH and salinity. These factors are presented in the following sections.

2.1.1 Fluorescence quenching

Fluorescence quenching is a term, which covers any process that leads to a decrease in fluorescence intensity of a sample. It is a deactivation of the excited molecule either by intra- or intermolecular interactions. Quenching can be divided into two main categories: static and dynamic quenching.

When the environmental influence (quencher) inhibits the excited state formation, the process is referred to as static quenching. Static quenching is caused by ground state complex formation, where the fluorophore forms non-fluorescent complexes with a quencher molecule. Dynamic quenching or collisional quenching refers to the process when a quencher (e. g. oxygen) interferes with the behaviour of the excited state after its formation. The excited molecule will be deactivated by contact with other molecules or by intermolecular interactions (collision). A wide variety of substances can act as quenchers of fluorescence for different fluorophores (Christensen, 2005; Lakowicz, 2006). In table 1, the quenchers of typical fluorphores are presented.

Typical fluorophore(s)	Quencher(s)
Tryptophan	Acrylamide, halogen anesthetics, hydrogen peroxide, imidazole, histidine, picolinium nicotinamide, succinimide, trifluoroacetamide
Anthracene	Amines, halogens, iodide, thiocyanate
Tyrosine	Disulfides
Polycyclic aromatic hydrocarbons	Nitromethane and nitro compounds
Aromatic hydrocarbons, chlorophyll	Quinones
Naphthalene	Nitroxides, nitric oxide, halogens
Most fluorophores	Oxygen

Table 1. Fluorescence quenchers of typical fluorophores (adapted from Lakowicz, 2006).

Generally, in water the most important fluorescence quencher with high impact on the fluorescence response is temperature. Quenching is enhanced with increasing temperature determining the electrons within a molecule to return to the ground state by a radiationless process. In a study on dissolved organic matter (DOM) thermal fluorescence quenching, Baker (2005) showed that by decreasing the temperature from 45^0 C to 10^0 C the DOM fluorescence intensity increased with ~ 48 %. According to Baker's study (2005) the most affected fluorophore is tryptophan in comparison with fulvic acid.

Fluorescence quenching of dissolved organic matter (DOM) can also be induced using certain metal ions, like Cu^{2+}, Fe^{2+}/Fe^{3+}, Al^{3+}, etc. by the process of complex formation. Metal quenching affects mostly the humic substances and less the amino acids. Most studies have been performed in laboratories, under controlled conditions and little is known about the effects on natural organic matter (Kelton et al., 2007; Reynolds & Ahmad,1995).

2.1.2 Concentration and inner filter effect

Within the context of fluorescence measurements, the inner filtering effect (IFE) represents an apparent decrease in emission quantum yield and/or a distortion of band shape as a result of the absorption and emitted radiation by the sample matrix (Henderson et al, 2009). The fluorescence intensity is attenuated by:

- **Primary inner-filter effect,** referring to the absorption of the excitation beam prior to reaching the interrogation zone;
- **Secondary inner filter effect,** which refers to the absorption of the emitted fluorescence photons (Ohno, 2002);
- **Inner filter effects due to the presence of other substances.** When the solution contains other chromophores that absorb in the same wavelength range as the fluorescent compound under study, the chromophores act as filters at the excitation wavelength and the fluorescence intensity must be multiplied by a correction factor (Valeur, 2001).

In order to be easily understood, the primary and secondary IFE are graphically presented in figure 2.

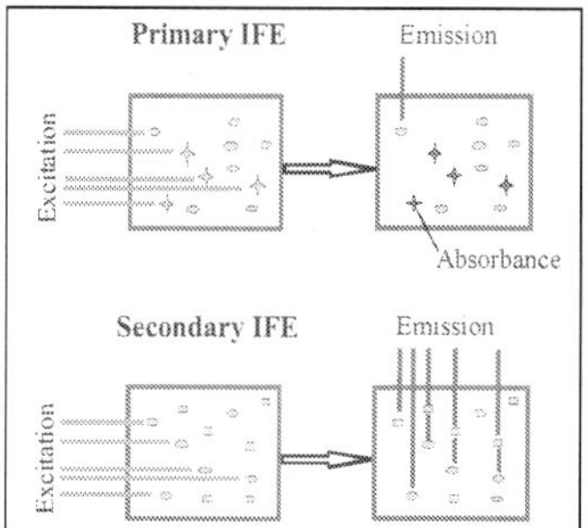

Fig. 2. Diagram of the mechanisms of inner filter effect: primary IFE and secondary IFE.

Various authors have suggested different approaches to correct for IFE, the two most common being an empirical correction based on the Raman scatter peak and a mathematical one based on absorbance profile of the same sample (Parker and Barnes, 1957; Lakowicz, 2006). An alternative approach is to leave the data uncorrected, and utilise the resulting wavelength-dependent non-linear relationship between fluorescence intensity and concentration (Henderson et al., 2009).

The IFE can also be avoided by using front-face illumination because it offers the advantage of being much less sensitive to the excitation inner filter effect. The illuminated surface is better oriented at 30^0 than at 45^0, because at 45^0 the unabsorbed incident light is partially reflected towards the detection system, which may increase the stray light interfering with the fluorescence signal (Valeur, 2001).

Another technique to minimise the IFE is to reduce the path length of the excitation light through the cuvette, but only the primary IFE is reduced. Simple sample dilution to a concentration at which IFE effects are negligible has also been suggested (Baker and Curry, 2004; Baker et al., 2004). There have also been recommendations towards the appropriate concentration quantified as absorbance values. Kubista et al. (1994) suggested that IFE does not occur if the samples show absorbance values lower than 0.05, while Pagano and Kenny (1999) indicate a threshold of 0.01.

2.1.3 Influence of pH on the fluorescence

The pH value of the sample affects the fluorescence of a fluorophore. The pH influence on fluorescence intensity of DOM components always presents the same trend: intensities increase with higher pH until 10, as observed by Reynolds and Ahmad (1995) at raw sewage samples. A small plateau is seen at pH from 5 to 7. Generally, it is not recommended to alter the pH of a sample, but special attention should be paid and the fluorescence spectra should be corrected (Baker, 2007). Similar results have been obtained by Patel-Sorentino et al. (2002) for the humic substances, only that after a fluorescence intensity increase until 10, a slight decrease of its intensity occurs at pH 12.

According to Patel-Sorentino et al. (2002), there are three possible hypotheses:

* **Alteration of the molecular orbital of the excitable electrons**, as a consequence of ionization of the fluorescent molecules after modifications of pH.

* **Macromolecular configuration of humic substances**: the more rigid structures are giving better fluorescent yields. Ghosh and Schnitzer (1980) observed that the structure of humic substances varied with pH changes. Their conclusion is that humic substances have linear structure at high pH and a coil one when pH decreases. Patel-Sorentino et al. (2002) also explained that a spherocolloied configuration could mask some fluorophores inside their structure. At higher pH, the configuration becomes linear and some fluorophores, which are not anymore masked, can fluoresce, increasing the fluorescence intensity.

* **Metal ions present in freshwaters**. This implies that there are some competition phenomena between H^+ ions and metal ions to complex DOM in freshwater, leading to complexation-decomplexation processes which directly affect fluorescence intensity. However, Patel-Sorentino et al. (2002), also note that the metal ions concentration in freshwater samples would be lower than the concentration of metals that quench the fluorescence of DOM. Therefore, the metal ions, which can increase the fluorescence, would have a too weak effect to produce a significant variation in DOM fluorescence intensity.

2.1.4 Salinity influence on the fluorescence

Salinity can affect DOM fluorescence by altering intramolecular reactions, such as conformational change and charge transfer. This results in an increased photoreactivity and fluorescence loss in certain fluorescence compounds (Osburn, 2001; Chen et al., 2002). The relationship between salinity and fluorescence intensity could help detect the source of natural organic matter in marine waters (Elliot, 2006; PhD Thesis).

Del Castillo et al. (1999) investigated changes in chromophoric DOM composition by studying the shifts in the fluorescence maxima. Where the salinity at a site was high, the position of the emission maximum at 350 nm excitation wavelength was shifted to shorter wavelengths, suggesting that high salinity leads to changes in chromophoric DOM.

As shown in previous sections, fluorescence technique is a powerful tool in analysing different samples, but several environmental factors must be taken into account. Most problems arise at highly polluted samples, which imply high concentration of the contaminant and in this case filtration, dilution and absorption to check for IFE are recommended. The pH, salinity and temperature of the sample should be measured, if possible, and the excitation and emission spectra should be corrected, if parameters values are above or under the normal domain. No correction is needed if the sample temperature, at the time of measurement, is between 20^0 C and 25^0 C, or if the pH is between 6 and 8. In conclusion, the general recommendation is that all these parameters need to be measured before the fluorescence analysis are performed and reported in the scientific literature.

2.2 Techniques for fluorescence spectra recording

The fluorescence signal is typically recorded as a: fluorescence emission spectrum, fluorescence excitation spectrum, synchronous fluorescence spectrum, total synchronous fluorescence spectrum or excitation – emission spectrum (Figure 3). An emission spectrum consists in the wavelength distribution of the light emission, measured at a single constant excitation wavelength (Figure 3a). Conversely, an excitation spectrum represents the dependence of emission intensity, measured at a single emission wavelength, upon the excitation wavelengths.

In most cases, the analysed samples contain complex multi-component mixtures which cannot be resolved satisfactorily by conventional fluorescence methods. Due to these gaps, for rapid, sensitive, and selective fluorescence analysis, three state-of–the-art methods have been introduced: synchronous fluorescence spectroscopy (SFS), total synchronous fluorescence spectroscopy (TSFS) and excitation-emission matrix (EEM) (Coble, 1996, Deepa and Mishra, 2006; Hudson et al., 2007). As mentioned earlier, an emission or excitation spectrum is recorded by separately scanning the excitation, respectively emission monochromator at various wavelengths. SFS spectra are recorded by scanning both monochromators simultaneously (Deepa and Mishra, 2006) (Figure 3b). Using SFS, the spectral band is narrowed and sharper peaks can be obtained by applying the optimum wavelength offset ($\Delta\lambda$) between excitation and emission. A SFS spectrum is illustrated as fluorescence intensity function of excitation wavelength, for a certain $\Delta\lambda$ (Figure 3c). Total synchronous fluorescence spectrum offers more selectivity and sensitivity to multi-fluorophores mixture analysis. It is presented as a contour map, containing numerous synchronous spectra at different offsets gathered into one bi-dimensional image (Figure 3c). An example of TSFS map is shown in figure 1.18 for a water sample with high protein-like fluorescence intensity.

The last method for complex multi-compounds mixture detection is to record the fluorescence signal as excitation-emission matrices. EEMs represent fluorescence contour maps, in which repeated emission scans are collected at numerous excitation wavelengths providing highly detailed information (Coble, 1996; and references therein) (Figure 3d). Coble (1996) mentions that, once the EEMs have been fully corrected for instrumental configuration, data can be analyzed as excitation spectra, emission spectra or surface spectra, even though originally collected as emission scans (Figure 3d). The EEMs are very simple to analyse because the fluorescence intensity maximum are identified as $\lambda_{excitation}$ / $\lambda_{emission}$ pairs. Usually, the images are colour coded, the highest intensity being represented with red and the lowest with blue.

At a closer inspection of a water fluorescence spectrum, other maxima can be observed. These belong to the scattering of the incident light and are most intense when dealing with turbid solutions and solid opaque samples. Scattering can affect the fluorescence signal, therefore it is of utmost importance to check the absorbance measurements and correct the fluorescence response. Scattered light can be divided into Rayleigh scatter and Raman scatter, according to its nature. Rayleigh scatter is the scattering of light by particles and molecules smaller than the wavelength of the light. Rayleigh scattering represents so-called elastic scatter, meaning that no energy loss is involved, so that the wavelength of the scattered light is the same as that of the incident light. The Rayleigh scatter can be observed as a diagonal line in fluorescence landscapes for excitation wavelengths equalling the emission wavelengths, as seen in figure 3d. Due to the construction of grating monochromators used for excitation in most spectrofluorometers, also some light at the double wavelength of the chosen excitation will pass through to the sample. For this reason an extra band of Rayleigh scatter, 2nd order Rayleigh, will typically appear in fluorescence measurement (Christensen, 2005).

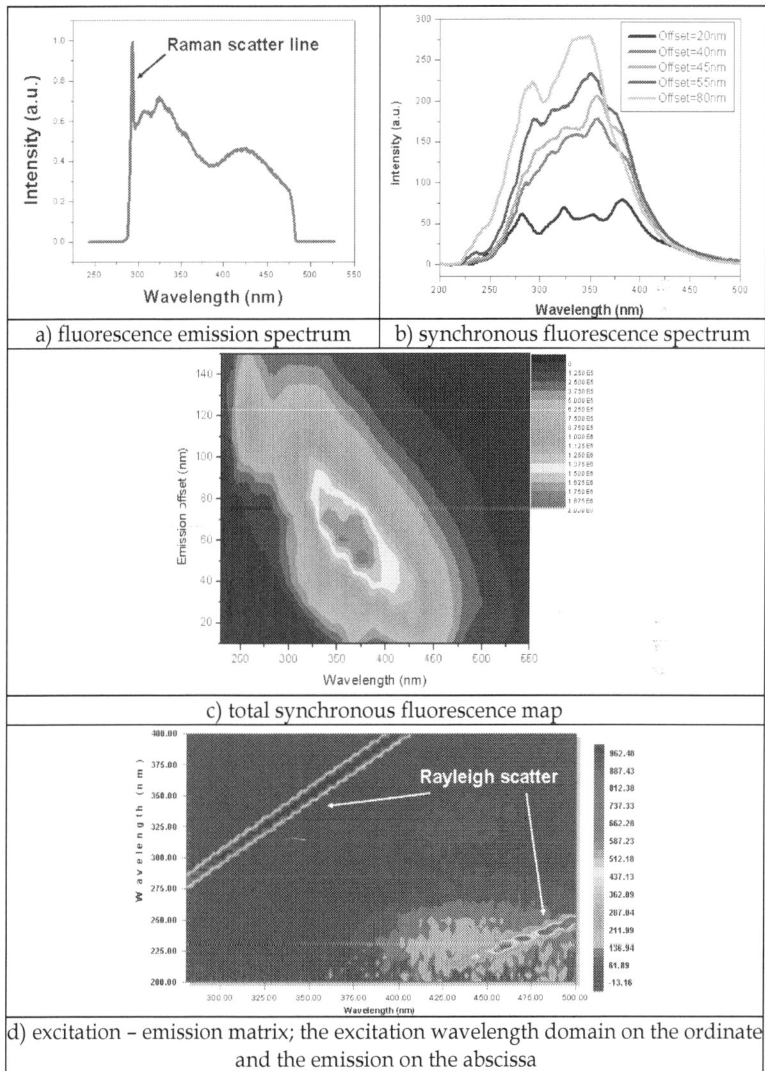

Fig. 3. Typically recorded fluorescence spectra for water samples

Raman scatter is an inelastic scatter, caused by the absorption and re-emission of light coupled with vibrational states. A constant energy loss will appear for Raman scatter, meaning that the scattered light will have a higher wavelength than the excitation light, with a constant difference in wavenumbers. In figure figure 3a, the Raman scattering can be seen as a diagonal line with a systematic, increasing deviation from the Rayleigh scatter line, since the axis is shown as wavelengths, which is not proportional to the energy of the light. The Raman scatter line can be used to check for instrument stability and to quantify the degree of contamination from a water sample by using the normalised fluorescence intensity to the Raman peak. The advantages offered by the Raman line are: (a) the independence of the chemistry since it measures the properties of the solvent; (b) the ease of application and sensitivity; (c) versatility since it can be applied at any wavelength between 200 and 500 nm. In the case of water the Raman line offers the advantage that it is very stable, appearing in the spectrum at the same offset from the excitation wavelength.

3. Dissolved organic matter fluorescence

The dissolved organic matter (DOM), the ubiquitous fraction in soil and aquatic ecosystems, is a heterogeneous mixture of humic substances, fatty acids and phenolic compounds, amino acids, nucleic acids, carbohydrates, hydrocarbons and other compounds, being among the largest reservoirs of carbon on the planet (Spitzy and Leenheer, 1991; Thomas, 1997; Swietlik and Sikorska, 2004). The dynamics and characteristics of DOM strongly influence a number of key ecosystem processes, including the attenuation of solar radiation, control of nutrient availability, alteration of contaminant toxicity, material and energy cycling (Cammack 2002; PhD Thesis; and references therein). The composition of DOM differs depending on source: it is estimated to contain 0.5 mg/L dissolved organic carbon in alpine streams or 100 mg/L in wetland streams (Spitzy and Leenheer, 1991; Frimmel, 1998). Only 25 % of DOM is fully characterized. It is estimated that 40–70 % from aquatic dissolved organic matter is composed of humic substances (Thurman, 1985; Senesi, 1993).

By the type of production, DOM can be classified as natural or derived from human activity (human wastes, farm wastes, leachates, etc.), but by the origin, DOM can be either allochthonous or autochthonous. Allochthonous DOM, is the fraction that is formed outside the water system and transported inside through discharge, geological and land-use activities or dry and wet deposition (McDowell and Likens, 1988; Hudson et al., 2007). The composition and concentration of allochthonous DOM, in aquatic systems, is dependent mostly on the soil type, catchment, precipitation, vegetation, flow path of water through different soil horizons and other soil processes (Hope et al., 1997; Aitkenhead et al., 1999). Autochthonous DOM is formed within the water system, through derivation from polymerisation and degradation of existing DOM, release from living and dead organisms and through microbial syntheses within the body water (Thomas, 1997).

Within the complex heterogeneous mixture of DOM, only the following components are mostly studied by fluorescence: proteins and humic substances. The protein fluorescence is given by the amino acids tryptophan, tyrosine and phenylalanine, and is related to the activity of bacterial communities, as shown by Cammack et al. (2004) and Elliot et al. (2006a). The humic substances fluorescence indicates the break-down of plant material by biological and chemical processes in the terrestrial and aquatic environments (Elkins and Nelson, 2001; Stedmon et al., 2003; Patel-Sorrentino et al., 2004). Humic substances are divided into two major fractions, depending on the solubility at different pH values: humic acids which are insoluble in aqueous solution at pH lower than 2, but soluble at higher pH and fulvic acids soluble in water under all pH conditions (Aiken et al., 1985). DOM fluorophores are schematically represented in figure 4, along with their corresponding excitation/emission wavelengths domains. Due to the difficulties associated with identifying of the individual fluorescent compounds in waters, these groups of fluorophores are commonly named humic-like, fulvic-like and protein-like (specifically tryptophan- or tyrosine-like), so called because their fluorescence occurs in the same area of optical space as the standards of these materials (Hudson et al., 2007).

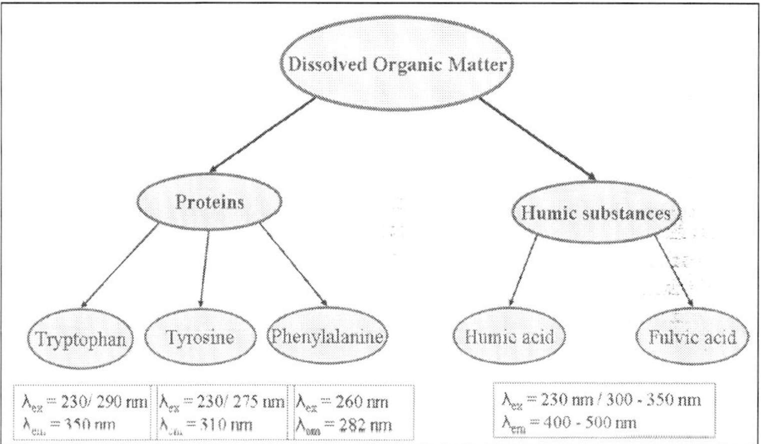

Fig. 4. Schematic representation of DOM fluorescent fractions with the specific excitation/emission wavelengths domains.

Beside standard nomenclature (e.g. humic-like), Coble (1996) defined the humic substances as peak **A** ($\lambda_{excitation}$ = 230 nm, $\lambda_{emission}$ = 400 – 500 nm) and peak **C** ($\lambda_{excitation}$ = 300 – 350 nm, $\lambda_{emission}$ = 400 – 500 nm), tryptophan as peak **T** and tyrosine as peak **B**. Tryptophan, also, presents two excitation wavelengths, therefore, T_1 corresponds to the peak at $\lambda_{excitation}$ = 290 nm and T_2 to the peak at $\lambda_{excitation}$ = 230 nm (Figure 5).

Certain types of contaminants and their relative impact on the system can only be determined by analysing the fluorescence intensity, excitation and emission wavelengths of the above mentioned fluorophores. In the past decades, numerous studies have shown that these fluorophores can provide more information about the characteristics of DOM and the aquatic system. According to some studies, peak C fluorescence intensity correlates with total organic carbon (Smart et al., 1976; Vodacek et al., 1995; Ferrari et al., 1996) and shows a linear relationship with aromaticity (McKnight et al., 2001). Fluorescence intensity of peak C also relates with the molecular weight of the organic fractions, showing lower values for smaller molecular weight fractions (Stewart and Wetzel, 1980). Peak C emission wavelength shows the degree of hydrophobicity, a higher emission wavelength corresponding to greater degree of hydrophobicity (Wu et al., 2003). Peak T presents a very strong correlation with the standard parameter biological oxygen demand (BOD). Some researchers (Reynolds and Ahmad, 1999; Hudson et al., 2008) even tested the possibility of using peak T fluorescence as a surrogate for the standard water quality parameter, BOD. The relationship between biological activity of aquatic plankton, along with algae metabolism rates, and peak T fluorescence intensity for different DOM types has also been observed (Bieroza et al., 2009 and references therein).

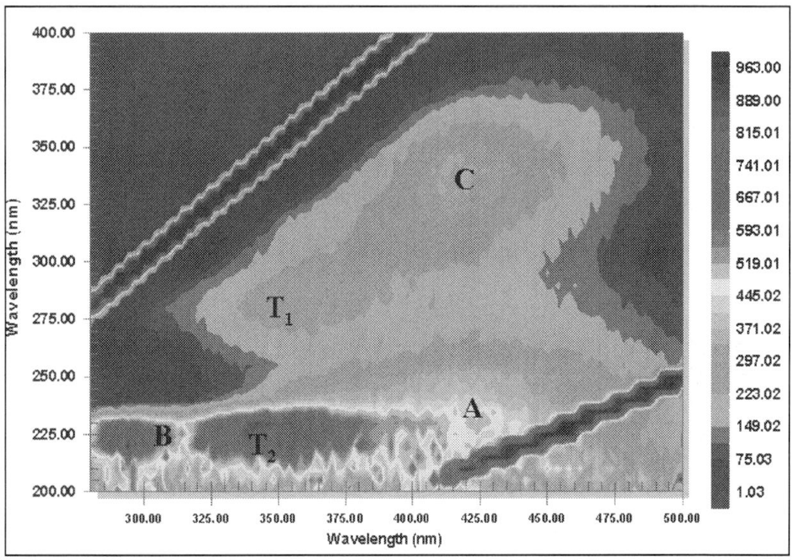

Fig. 5. Excitation – emission matrix presenting the fluorescence domains of the humic – like, peaks A and C, and protein-like fractions, peak B (tyrosine) and peaks T_1 and T_2 (tryptophan).

4. Fluorescence properties of common aquatic pollutants

Water systems naturally contain, as shown in previous sections, organic matter with two very important fluorescence components: humic substances and proteins, which differ in quantity depending on the water body. When high quantities of one component is produced or released into the water, then the balance of the ecosystem is disrupted with potentially long-term effects. Apart from humic substances and proteins, there are other compounds that can contaminate the water and can be detected with fluorescence spectroscopy: polycyclic aromatic hydrocarbons (PAHs), pesticides, environmental hormones. Pesticides and PAHs reach the aquatic environment through direct runoff, leaching, careless disposal of empty containers, equipment washing a.s.o. (Konstantinou et al., 2006). Due to their toxic nature, persistence in the environment and presence in any type of system (soil, surface water, groundwater) many studies concentrated on sensitive, selective and early detection of these pollutants (e.g. Ferrer et al., 1998; Jiji et al., 1999; Jiji et al., 2000; Selli et al., 2004; Deepa et al., 2008). The standard techniques for pesticides and PAHs detection are gas and liquid chromatography, which require tedious extraction or separations procedures and expensive equipments. Fluorescence spectroscopy is a rapid and cost effective alternative, since PAHs and many pesticides are naturally fluorescent (Jiji et al., 1999). When dealing with more components, the use of SFS or EEM techniques is recommended.

For instance, petroleum products contain a complex mixture of generically classified as: aromatic hydrocarbons and alyphatic hydrocarbons. Only aromatic hydrocarbons exhibit fluorescence and the emission wavelength is proportional to the number of aromatic ring one compound has. Therefore, monoaromatic compounds (benzene, toluene, xylene and phenols) emit fluorescence between 250 – 290 nm. Two aromatic ring compounds, like naphthalene, show a fluorescence peak at 310 – 330 nm, phenanthrenes (three aromatic rings) between 345 – 355 nm and so on (Pharr et al., 1992; Abbas et al., 2006). Pharr et al. (1992) has shown that, only by using standard fluorescence emission spectra, distinguishing between 2 brands of gasoline would be impossible.

Care should be taken when measuring petroleum products if sophisticated fluorescence spectra are recorded. Since they contain a large number of fluorophores, it is difficult to choose the proper excitation domain and if there is a high concentration of petroleum products in the sample the inner filter effect can interfere and change the real feature of fluorescence emission (Ryder, 2005). The phenomenon of concentration-dependent red-shift of fluorescence, observed in multifluorophoric systems at high concentrations, has been successfully used in the analytical fluorimetry for systems like petroleum derivatives, humic substances, biological fluids, etc. Divya and Mishra (2008) proposed a method to get the appropriate excitation wavelength by using derived absorption spectra for different concentration of petroleum products (Figure 2.6).

The spectral fingerprints for the major petroleum products that can be present as environmental pollutants (petrol, kerosene, Diesel and engine oil) have been acquired by Patra and Mishra (2002a), Cristescu et al. (2009) and Carstea et al. (2009a). The authors observed that heavy oil, Diesel and engine oil, show fluorescence maxima in the longer emission wavelength region (420 – 550 nm) while lighter oils, petrol and kerosene, present peaks mainly in an intermediate wavelength region (310 – 400 nm).

Patra and Mishra (2002b) also studied the effects of adulteration of petroleum products on the fluorescence signal. The authors found that the emission wavelength changes according to the concentration of adulterant into the solution. Similar effect has been observed at petrol adulterated with kerosene, in a more recent study performed by Divya and Mishra (2008).

Beside adulteration, the effects of age on petroleum products have been studied (Li et al., 2004; Deepa et al., 2006). It is well-known that the persistence of PAHs in the environment depends on numerous factors as: physical and chemical characteristics of PAHs and medium, concentration, dispersion and bioavailability of PAHs. Generally, high molecular weight PAHs (> 4 rings) present higher degree of toxicity and longer persistence in the environment compared to low molecular weight PAHs (< 3 rings). For example, the tricyclic phenantherene half-life ranges from 16 to 126 days in soil, whereas benzo[a]pyrene (5 rings) has a half-life ranging from 229 to 1,500 days (Chauhan et al., 2008). According to this fact, one can assume that the fluorescence spectrum could change if a petroleum product ages. Deepa et al. (2006) studied the fluorescence signal of transformer oil during its aging process, which was thermally induced at 100°C for 31 days. They observed a sudden dramatic decrease in fluorescence intensity after only 17 days, for raw oil, followed by a slight increase until day 31. Also, the excitation and emission maxima increased starting with the 20th day. Deepa et al. (2006) explained that when transformer oil was degraded, its acidity increased, resulting an increase in C-O band and C=C double bands. During thermal decomposition, paraffinic compounds have dehydrogenated and formed hydroperoxides, resulting in, after oxidation, aldehydes and ketones by a free radical mechanism. The authors explained that the sudden drop in fluorescence intensity, on the 17th day, could be caused by the presence of antioxidants in the oil sample which inhibited degradation until the antioxidants were consumed.

Li et al. (2004) evaluated the fluorescence signal of more aged products, but using a different method for adulteration, compared to Deepa et al. (2006). Weathering was induced after preserving the samples in a refrigerator for no more then 15 days. The researchers concluded that, because the weathering process is so complex and unpredictable, the attempts to both characterize the oil and the exact extent of weathering would be impossible. At one sample, which presented, before weathering 3 emission peaks at ~ 360, 375 and 415 nm, when excited with 254 nm, a decrease in fluorescence intensity was noticed. PAHs can enter and pollute the environment not only by petroleum products, but also by car tyres, coal tar or creosote. Therefore the analysis of distinct PAHs is important, especially for the 16 PAHs included in the EPA list as being very toxic, mutagenic and carcinogenic. Giamarchi et al. (2000) recorded the fluorescence spectra for 5 PAHs, most commonly tested in drinking water: fluorene, naphthalene, phenantherene, benzo[a]pyrene, fluoranthene, with excitation wavelength at 263 nm, illustrated in figure 2.9. Also, Giamarchi et al. (2000) obtained almost the same emission peaks in a mixture of the 5 PAHs, both at high concentration and low concentration. Naphthalene fluorescence was overlapped by fluorene due to high intensity and one peak belonging to benzo[a]pyrene and phenanthrene was also overlapped. Ferrer et al. (1998) applied SFS to 10 PAHs, as pure samples and mixtures and obtained similar results. Hence, Giamarchi et al. (2000) recommend the separation of PAHs, from the mixture, in order to clearly identify each component. Similar study has shown that by using a mathematical model the compounds can be separated. Jiji et al. (1999) resolved the spectra of pyrene and chrysene, from a mixture, with 3 way-PARAFAC model. In the same way, Jiji et al. (1999) separated the spectra of pesticides which also contain a mixture of fluorophores which overlap. Some authors (Burel-Deschamps et al., 2006; Pascu et al., 2001) used absorption spectroscopy and laser induced fluorescence for pesticide monitoring in water. Organochlorurate pesticides in water, crude oil and oil components in water and soil with detection limits of 10^{-1}–10^{-2} ppm were obtained.

In conclusion, fluorescence spectroscopy can be used not only to detect protein-like and humic-like fractions, but also petroleum and pesticide pollution. It is highly important, when dealing with these pollutants, to identify the specific contaminant in order to act quickly for decontamination. But, fluorescence spectroscopy can only be applied for preliminary information about a contaminant, researchers recommending subsequent standard analyses.

5. Real-time monitoring of water quality

Fluorescence spectroscopy has been intensely used in recent decades for the analysis of DOM and organic pollutants in water. Additionally, fluorescence technique correlates with standard parameters, like biological oxygen demand or dissolved oxygen (Pfeiffer et al., 2007; Hudson et al., 2008). Some researchers have pointed out that it may be possible to use fluorescence spectroscopy for water quality monitoring purposes, in order to identify DOM characteristics at temporal scale, detect pollutants and be used as a surrogate for standard measurements (for example, Ahmad & Reynolds, 1999; Henderson et al., 2009; Carstea et al. 2010).

5.1 Fluorescence fingerprints of various water types

In order to identify the type of pollution in a water system, it is necessary to establish the natural characteristics of the water body, the quantity and quality of DOM and the relative proportions of terrestrially and microbially derived components. This can be obtained by determining the fluorescence fingerprint of a specific water body. Several studies had identified the specific fluorescence signature of different aquatic systems. Most of research concentrated on characterizing marine DOM properties, but few scientists attempted to analyse other water systems, like riverine, lentic systems, canals or surface runoff. Until approximately two decades ago, most researchers failed to distinguish between marine and freshwater DOM components, due to instruments limitations. Technological developments had a major contribution to fluorescence studies by allowing scientists to record complex two or three-dimensional spectra.

Coble (1990) and Coble et al. (1996) were the first ones to offer a comprehensive investigation of DOM fractions in marine and freshwaters. They found that the position of fluorescence peaks occur at shorter wavelengths for marine water than for freshwater. Also, a very intense peak B, corresponding to tyrosine, was seen at gulf and bay samples, which was not observed at riverine samples. Similarly, Kowalczuk et al. (2005) obtained different spectral characteristics at samples taken from estuarine, coastal and Gulf Stream waters. The researchers had found that the dominant component in estuarine spectra is the humic-like fluorescence and in the Gulf Stream is the protein-like. But, the coastal sample presented a mixture of these two fractions, due to significant microbial reprocessing and local production of DOM. Coble (1996) and Parlanti et al. (2000) had identified another peak, which can be seen only at marine samples, named as peak M, related to biological activity in areas of primary productivity, i.e. "fresh" humic material. Considering the variations in estuarine, coastal and marine water samples, Tedetti et al 2010 has evidenced that fluorescence spectroscopy is a very good tool for tracking anthropogenic inputs in the coastal waters.

Riverine samples have been analysed in few occasions, by scientists, in order to identify the peculiarities of a certain water body at temporal and spatial scale (Hudson et al., 2007). Most studies had shown that the humic substances largely dominate the fluorescence spectra, but the protein-like fraction can also be detected, depending on the type, location and inputs of the river (Baker et al., 2004; Carstea et al., 2009a; Carstea et al., 2009b). It has been shown that riverine DOM presents a seasonal variation (Baker et al., 2003; Carstea unpublished data), but also a subtle daily variation (Spencer et al., 2007). Riverine DOM varies greatly at spatial scale, as evidenced for example by Baker & Spencer (2004), who have undertaken a study on samples collected from several locations along a river, from source to its flow into the sea. Researchers have proved that fluorescence emission wavelengths of humic substances tend to decrease with distance from the source of the river, while the fluorescence intensity increased. Fluorescence intensity was highly influence by the type of surrounding (rural or urban) and activity (agriculture, airport etc.).

Even more sensitive to the surrounding environment are the lentic systems, as evidenced by Ghervase et al. (2011). The authors state that fluorescence spectra are dominated by the presence of the protein-like component of allochthonous origine, due to the high quantities of surface runoff or due to fauna and vegetation within the ecosystem. Borisover et al. (2009) has identified both DOM components in a lake with multiple river inputs and a seasonal variation at the surface of the lake.

Below, examples of fluorescence EEMs for a rural river, urban river, lake and a sewage-impacted river are presented in order to better illustrate the difference in fluorescence properties of DOM (Figure 6). Rural sample fluorescence spectrum shows both DOM fractions, originating from farm wastes, while the urban sample presents mostly the humic-like component. At the lake sample, high intensity humic-like component can be seen. The sample with sewage influence shows great impact from pollution by intense peaks corresponding to the protein-like fraction. More details regarding fluorescence fingerprints of various aquatic systems can be found in (Carstea et al., 2009b).

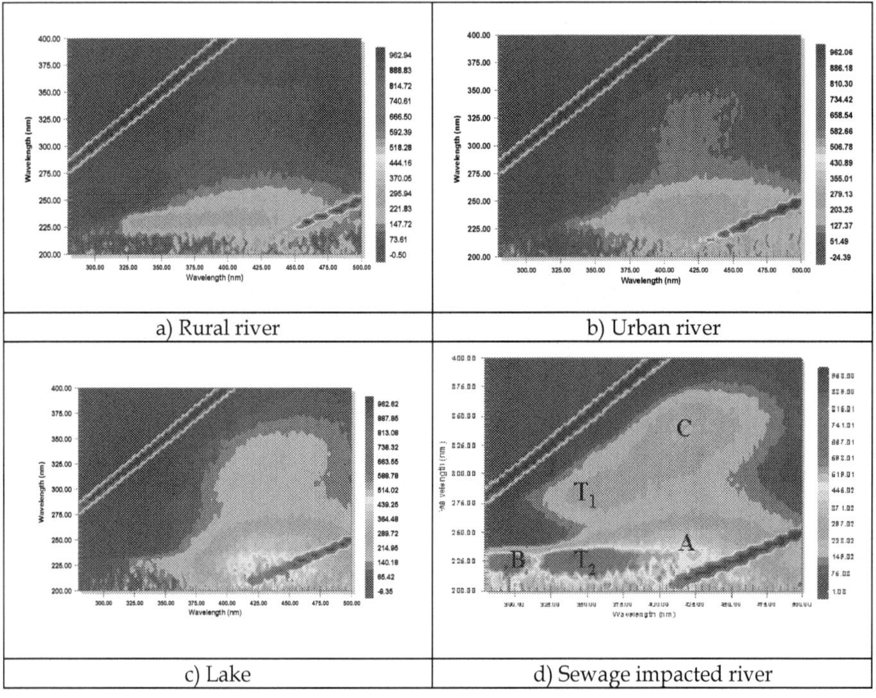

a) Rural river	b) Urban river
c) Lake	d) Sewage impacted river

Fig. 6. Fluorescence excitation –emission matrix for water samples.

5.2 Fluorescence monitoring of freshwater systems

So far, real-time fluorescence data have mostly been obtained for marine water with instruments specifically designed for this application (Chen, 1999; Barbini et al., 2003; Conmy et al., 2004; Drozdowska, 2007). One of the techniques used for real-time analyses is laser induced fluorescence (LIF). Chen (1999), Drozdowska (2007) and Bukin et al. (2007) have used LIF, at various wavelengths, to characterize DOM, arriving at the conclusion that with fluorescence rapid mapping of major processes can be made without sampling artifacts and measurement delays. Belzile et al. (2005) and Downing et al. (2009) have used a WebStar submersible fluorometer ($\lambda_{ex}/ \lambda_{em}$ = 370 / 460 nm), while Del Castillo et al. (2001) and Conmy et al. (2004) used a SAFIre model (multiple excitation filters from 228 – 490 nm and emission filters at 228 – 810 nm). They reached at the conclusion that fluorescence spectroscopy is highly capable to assess in real-time DOM characteristics and detect small differences in optical properties. However, the calibration of the system is time consuming and the excitation and emission domains are very limited.

Similar equipment has been used for freshwater studies in order to obtain continuous real-time data (Spencer et al., 2007; Downing et al., 2009). Spencer et al. (2007) has undertaken the first in situ monitoring of DOM properties, specifically the humic component, at temporal scale. The authors were able identify a diurnal variation of DOM using fluorescence spectroscopy, but recommended the analysis of multiple optical parameters in order to capture full characterization of DOM variability.

However, the in situ DOM monitoring was limited by the fixed excitation and emission wavelengths, allowing the measurement of only the terrestrial components. Another issue was the clogging of pre-filters, which required frequent filter replacement. Continuous monitoring has been performed by Carstea et al. (2009) who collected water samples at hourly scale, but the measurements were made within 24 h and not in real-time. This was the first research of continuous monitoring on fluorescence EEMs, allowing assessment of both DOM fractions. These studies revealed that dissolved organic matter varied at a daily timescale, depending on the river type, and was highly influenced by precipitation. As this study raised many topics for further research, Carstea et al. (2010) attempted to undertake the first study on real-time analysis of DOM using fluorescence EEMs. The experiment was made on a small urban catchment using a standard bench-top fluorometer connected to a fibre-optic probe. Researchers have proved that fluorescence technique is very reliable for real-time studies and that it can be applied in several systems for at least two week, without any cleaning procedure of the tubing system. Hourly pollution pulses were detected together with a significant contamination event with diesel oil. In conclusion, fluorescence spectroscopy is a very effective technique for in situ monitoring, offering many possibilities for further research for various applications.

5.3 Potential applications

It has been established, so far, that this technique can characterize natural organic matter (Baker and Spencer, 2004; Winter et al., 2007) and detect different types of aquatic pollutants, like sewage (Baker, 2001; Reynolds, 2002), oil (Budgen et al., 2008; Patra and Mishra, 2002; Carstea et al., 2010) or pesticides (Jiji et al., 2000). Furthermore, DOM fluorescence data correlate with water quality parameters such as: total organic carbon (Vodacek et al., 1995), aquatic plankton (Mopper and Schultz, 1993), faecal coliforms (Pfeiffer et al., 2008) and biological oxygen demand (Reynolds and Ahmad, 1997; Hudson et al., 2008).

Although the technique advantages have been thoroughly tested, in various conditions, there are still multiple potential applications, which have not been tested, yet. For instance, Ahmad and Reynolds (1999) have suggested this method for on-line process control in sewage treatment plants. Subsequently, other studies have implied the use of fluorescence spectroscopy as a potential monitoring tool for recycled water (Henderson et al.,2009), drinking water treatment processes (Cheng et al., 2004; Bieroza et al., 2009), urban watersheds with sewage effluents (Hur et al., 2008) and evaluation of DOM composition and concentration in relation to the production of disinfection by products during drinking water chlorination (Spencer et al., 2007 and references therein). However, many issues have to be clarified before using this method for the previously mentioned application: possibility for automated data analysis, optimized calibration procedure to ensure reliability and repeatability of results and optimum instrument configuration (Henderson et al., 2009).

6. Conclusions

The study has presented fluorescence spectroscopy ability to characterize in real-time and in situ DOM fractions properties. Fluorescence spectroscopy can be influenced by numerous factors, but these issues can be overcome by the use of a calibration curve.

Fluorescence fingerprints are very useful in establishing the natural characteristics of the water body, the quantity and quality of DOM and the relative proportions of terrestrially and microbially derived components. Based on the natural characteristics, researchers can easily identify a pollution event. Fluorescence spectroscopy presents several opportunities for future research with potential application in drinking water and waste water treatment monitoring.

7. Acknowledgment

This work was supported by a grant of the Romanian National Authority for Scientific Research, CNCS – UEFISCDI, project number PN-II-RU-TE-2011-3-0077.

8. References

Abbas, O., Rebufa, C., Dupuya, N., Permanyer, A. & Kister, J. (2006). Assessing petroleum oils biodegradation by chemometric analysis of spectroscopic data. *Talanta*, Vol. 75, No. 4, pp. 857–871, ISSN 0039-9140.

Ahmad, S.R. & Reynolds, D.M. (1999). Monitoring of water quality using fluorescence technique: Prospect of on-line process control. *Water Research*, Vol. 33, No. 9, pp. 2069-2074, ISSN 0043-1354.

Aiken, G.R. (1985). Isolation and concentration techniques for aquatic humic substances, In: *Humic substances in soil, sediment and water: geochemistry and isolation*, Aiken, G.R., McKnight, D.M., Wershaw, R.L. & MacCarthy, P., pp. 363–385, Publisher Wiley-Interscience, ISBN 0471882747, New York.

Aitkenhead, J.A., Hope, D. & Billett, M.F. (1999). The relationship between dissolved organic carbon in stream water and soil organic carbon pools at different spatial scales. *Hydrological Processes*, Vol. 13, No. 8, pp. 1289-1302, ISSN 1099-1085.

Baker, A. (2001). Fluorescence excitation-emission matrix characterisation of some sewage impacted rivers. *Environmental Science & Technology*, Vol. 35, No. 5, pp. 948-953, ISSN 1520-5851.

Baker, A., Inverarity, R., Charlton, M. & Richmond, S. (2003). Detecting river pollution using fluorescence spectrophotometry: case studies from the Ouseburn, NE England. *Environmental Pollution*, Vol. 124, No. 1, pp. 57-70, ISSN 0269-7491.

Baker, A. & Curry, M. (2004). Fluorescence of leachates from three contrasting landfills. Water Research, Vol. 38, No. 10, pp. 2605-2613, ISSN 0043-1354.

Baker, A. & Spencer, R.G.M. (2004). Characterization of dissolved organic matter from source to sea using fluorescence and absorbance spectroscopy. *Science of the Total Environment*, Vol. 333, pp. 217-232, ISSN 0048-9697.

Baker, A., Ward, D., Lieten, S.H., Periera, R., Simpson, E.C. & Slater, M. (2004). Measurement of protein-like fluorescence in river and wastewater using a handheld spectrophotometer. *Water Research*, Vol. 38, No. 12, pp. 2934-2938, ISSN 0043-1354.

Baker, A. (2005). Thermal fluorescence quenching properties of dissolved organic matter. Water Research, Vol. 39, No. 18, pp 4405-4412, ISSN 0043-1354.

Baker, A., Elliott, S. & Lead, J.R. (2007). Effects of filtration and pH perturbation on organic matter fluorescence. Chemosphere, Vol. 67, No.10, pp. 2035-2043, ISSN 0045-6535.

Barbini, R., Colao, F., Fantoni, R., Ferrari, G.M., Lai, A. & Palucci, A. (2003). Application of a lidar fluorosensor system to the continuous and remote monitoring of the Southern Ocean and Antarctic Ross Sea: results collected during the XIII and XV Italian oceanographic campaigns. *International Journal of Remote Sensing*, Vol. 24, No. 16, pp. 3191–3204, ISSN 1366-5901.

Bieroza, M., Baker, A. & Bridgeman, J. (2009). Relating freshwater organic matter fluorescence to organic carbon removal efficiency in drinking water treatment. *Science of the Total Environment*, Vol. 407, No. 5, pp. 1765-1774, ISSN 0048-9697.

Budgen, J.B.C., Yeung, C.W., Kepkay, P.E. & Lee, K. (2008). Application of ultraviolet fluorometry and excitation–emission matrix spectroscopy (EEMS) to fingerprint oil and chemically dispersed oil in seawater. *Marine Pollution Bulletin*, Vol. 56, No. 4, pp. 677–685, ISSN 0025-326X.

Burel-Deschamps, L., Giamarchi, P., Stephan, L., Lijour, Y. & Le Bihan, A. (2006). Laser-Induced Fluorescence Detection of Carbamates Traces in Water. *Journal of Fluorescence*, Vol. 16, No. 2, pp. 177-183, ISSN 1573-4994.

Cammack, W.K.L. (2002). Dissolved organic matter fluorescence: Relationships with heterotrophic bacterial metabolism. PhD Thesis, Department of Biology, McGill University, Montréal, Québec, Canada.

Cammack, W.K.L., Kalf, J., Prairie, Y.T. & Smith, E.M. (2004). Fluorescent dissolved organic matter in lakes: Relationship with heterotrophic metabolism. *Limnology and Oceanography*, Vol. 49, No. 6, pp. 2034-2045, ISSN 1939-5590 .

Camobreco, V.J., Richards, B.K., Stenhuis, T., Peverly, J.H. & McBride, M.B. (1996). Movement of heavy metals through undisturbed and homogenized soil columns. *Soil Science*, Vol. 161, No. 11, pp. 740–750, ISSN 1538-9243.

Carstea, E.M., Baker, A., Boomer, I. & Pavelescu, G. (2009a). Continuous fluorescence assessment of organic matter variability on the Bournbrook River, Birmingham, UK. *Hydrological Processes*, Vol. 23, No. 13, pp. 1937–1946, ISSN: 1099-1085.

Carstea, E.M., Cristescu, L., Pavelescu, G. & Savastru D. (2009b). Assessment of the anthropogenic impact on water systems by fluorescence spectroscopy. *Environmental Engineering and Management Journal*, Vol. 8, No. 6, pp. 1321-1326, ISSN 1843-3707.

Carstea, E.M., Baker, A., Bieroza, M. & Reynolds D.M. (2010). Continuous fluorescence excitation emission matrix monitoring of river organic matter. *Water research*, Vol. 44, No. 18, pp. 5356-5366, ISSN 0043-1354.

Chauhan, A., Fazlurrahman, Oakeshott, J.G. & Jain, R.K. (2008). Bacterial metabolism of polycyclic aromatic hydrocarbons: strategies for bioremediation. *Indian Journal of Microbiology*, Vol. 48, No. 1, pp. 95–113, ISSN 0973-7715.

Chen, J., Gu, B.H., LeBoeuf, E.J., Pan, H.J. & Dai, S. (2002). Spectroscopic characterization of the structural and functional properties of natural organic matter fractions. *Chemosphere*, Vol. 48, No. 1, pp. 59-68, ISSN 0045-6535.

Chen, R.F. (1999). In situ fluorescence measurements in coastal waters. *Organic Geochemistry*, Vol. 30, No. 6, pp. 397-409, ISSN 0146-6380.

Cheng, W.P., Chi, F.H. & Yu, R.F. (2004). Evaluating the efficiency of coagulation in the removal of dissolved organic carbon from reservoir water using fluorescence and ultraviolet photometry. *Environmental Monitoring and Assessment*, Vol. 98, No. 1-3, pp. 421–431, ISSN 1573-2959.

Christensen, J. (2005). Autofluorescence of Intact Food - An Exploratory Multi-way Study. PhD thesis, Faculty of Life Sciences, University of Copenhagen, Denmark.

Coble, P.G., Green, S.A., Blough, N.V. & Gagosian, R.b. (1990). Characterization of dissolved organic matter in the Black Sea by fluorescence spectroscopy. *Nature*, Vol. 348, pp. 432 – 435.

Coble, P.G. (1996). Characterisation of marine and terrestrial dissolved organic matter in seawater using excitationemission matrix spectroscopy. *Marine Chemistry*, Vol. 51, No.4, pp. 325-346, ISSN 0304-4203.

Conmy, R.N., Coble, P.G. & Del Castillo, C.E. (2004). Calibration and performance of a new in situ multi-channel fluorometer for measurement of colored dissolved organic matter in the ocean. *Continental Shelf Research*, Vol. 24, No. 3, pp. 431–442, ISSN 0278-4343 .

Cristescu, L., Pavelescu, G., Carstea, E.M. & Savastru, D. (2009). Evaluation of petroleum contaminants in soil by fluorescence spectroscopy. *Environmental Engineering and Management Journal*, Vol. 8, No. 5, pp. 1269-1273, ISSN 1573-2959.

Deepa, S., Sarathi, R. & Mishra, A.K. (2006). Synchronous fluorescence and excitation emission characteristics of transformer oil ageing. *Talanta*, Vol. 70, No. 4, pp. 811-817, ISSN:0039-9140.

Deepa, S. & Mishra, A.K. (2006). Synchronous Fluorescence Spectroscopy and its Applications. *ISRAPS Bulletin*, Vol. 18, No. 1-2, pp. 4-8.

Del Castillo, C.E., Coble, P.G., Morell, J.M., Lopez, J.M. & Corredor, J.E. (1999). Analysis of the optical properties of the Orinoco River plume by absorption and fluorescence spectroscopy. *Marine Chemistry*, Vol. 66, No. 1-2, pp. 35-51, ISSN 0304-4203.

Divya, O. & Mishra, A.K. (2008). Understanding the concept of concentration-dependent red-shift in synchronous fluorescence spectra: Prediction of λ_{SFS}^{max} and optimization of $\Delta\lambda$ for synchronous fluorescence scan. *Analytica Chimica Acta*, Vol. 630, No. 1, pp. 47–56, ISSN:0003-2670.

Downing, B.D., Boss, E., Bergamaschi, B.A., Fleck, J.A., Lionberger, M.A., Ganju, N.K., Schoellhamer, D.H. & Fujii, R. (2009). Quantifying fluxes and characterizing compositional changes of dissolved organic matter in aquatic systems in situ using combined acoustic and optical measurements. *Limnology and Oceanography: Methods*, Vol. 7, pp. 119-131, ISSN 1939-5590.

Drozdowska, V. (2007). Seasonal and spatial variability of surface seawater fluorescence properties in the Baltic and Nordic Seas: results of lidar experiments. *Oceanologia*, Vol. 49, No. 1, pp. 59–69, ISSN 0078-3234.

Elkins, K.M. & Nelson, D.J. (2001). Fluorescence and FT-IR spectroscopic studies of Suwannee river fulvic acid complexation with aluminium, terbium and calcium. *Journal of Inorganic Biochemistry*, Vol. 87, No. 1-2, pp. 81-96, ISSN 0162-0134.

Elliott, S. (2006). Fluorescence signatures in natural aquatic systems. PhD thesis, University of Birmingham, UK.

Elliott, S., Lead, J.R. & Baker, A. (2006). Characterisation of the fluorescence from freshwater, planktonic bacteria. Water Research, Vol. 40, No. 10, pp. 2075-2083, ISSN 0043-1354.

Ferrari, G.M., Dowell, M.D., Grossi, S. & Targa C. (1996). Relationship between the optical properties of chromophoric dissolved organic matter and total concentration of dissolved organic carbon in the southern Baltic Sea region. *Marine Chemistry*, Vol. 55, No. 3-4, pp. 299-316, ISSN: 0304-4203.

Ferrer, R., Beltran, J.L. & Guiteras, J. (1998). Multivariate calibration applied to synchronous fluorescence spectrometry. Simultaneous determination of polycyclic aromatic hydrocarbons in water samples. *Talanta*, Vol. 45, No. 6, pp. 1073-1080, ISSN 0039-9140.

Frimmel, F.H. (1998). Characterization of natural organic matter as major constituents in aquatic systems. *Journal of Contaminant Hydrology*, Vol. 35, No. 1-3, pp. 201-216, ISSN 0169-7722.

Giamarchi, P., Stephan, L., Salomon, S. & Le Bihan, A. (2000). Multicomponent determination of a polyaromatic hydrocarbon mixture by direct fluorescence measurements. *Journal of Fluorescence*, Vol. 10, No. 4, pp. 393-402, ISSN 1573-4994.

Ghervase, L., Ioja, C., Carstea, E.M., Niculita, L., Savastru, D., Pavelescu, G. & Vanau G. (2011). Evaluation of lentic ecosystems from Bucharest City, *International Journal of Energy and Environment*, Vol. 5, No. 2, pp. 183-192, ISSN 1109-9577.

Ghosh, K. & Schnitzer, M. (1980). Macromolecular structures of humic substances. *Soil Science*, Vol. 129, pp. 266-276, ISSN 1538-9243.

Henderson, R.K., Baker, A., Murphy, K.R., Hambly, A., Stuetz, R.M. & Khan, S.J. (2009). Fluorescence as a potential monitoring tool for recycled water systems: A review. *Water Research*, Vol. 43, No. 4, pp. 863-881, ISSN 0043-1354.

Hope, D., Billett, M.F., Milne, R. & Brown, T.A.W. (1997). Exports of organic carbon in British rivers. *Hydrological Processes*, Vol. 11, No. 3, pp. 325-344, ISSN 1099-1085.

Hudson, N., Baker, A. & Reynolds, D. (2007). Fluorescence analysis of dissolved organic matter in natural, waste and polluted waters—a review. *River Research and Applications*, Vol. 23, No. 6, pp. 631–649, ISSN 1535-1467.

Hudson, N., Baker, A., Ward, D., Reynolds, D.M., Brunsdon, C., Carliell-Marquet, C. & Browning, S. (2008). Can fluorescence spectrometry be used as a surrogate for the biochemical oxygen demand (BOD) test in water quality assessment? An example from South West England. *Science of the Total Environment*, Vol. 391, No. 1, pp. 149–158, ISSN 0048-9697.

Hur, J., Hwang, S.J. & Shin, J.K. (2008). Using synchronous fluorescence technique as a water quality monitoring tool for an urban river. *Water, Air and Soil Pollution*, Vol. 191, No. 1-4, pp. 231–243, ISSN 1573-2932.

JiJi, R.D., Cooper, G.A. & Booksh, K.S. (1999). Excitation-emission matrix fluorescence based determination of carbamate pesticides and polycyclic aromatic hydrocarbons. *Analytica Chimica Acta*, Vol. 397, No. 1-3, pp. 61–72, ISSN: 0003-2670.

JiJi, R.D., Andersson, G.G. & Booksh, K.S. (2000). Application of PARAFAC for calibration with excitation–emission matrix fluorescence spectra of three classes of environmental pollutants. *Journal of Chemometrics*, Vol. 14, No. 3, pp. 171–185, ISSN 1099-128X.

Kelton, N., Molot, L.A. & Dillon, P.J. (2007). Spectrofluorometric properties of dissolved organic matter from Central and Southern Ontario streams and the influence of iron and irradiation. *Water Research*, Vol. 41, No. 3, pp. 638 – 646, ISSN 0043-1354.

Konstantinou, I.K., Hela, D.G. & Albanis, T.A. (2006). The status of pesticide pollution in surface waters (rivers and lakes) of Greece. Part I. Review on occurrence and levels. *Environmental Pollution*, Vol. 141, No. 3, pp. 555-570, ISSN 0269-7491.

Kubista, M., Sjoback, R., Eriksson, S. & Albinsson, B. (1994). Experimental correction for the inner-filter effect in fluorescence spectra. *The Analyst*, Vol. 119, No. 3, pp. 417–419.

Lakowicz, J.R. (2006). Principles of Fluorescence Spectroscopy. Third edition, Publisher Springer New York.

Li, J., Fuller, S., Cattle, J., Way, C.P. & Hibbert, D.B. (2004). Matching fluorescence spectra of oil spills with spectra from suspect sources. *Analytica Chimica Acta*, Vol. 514, No. 1, pp. 51–56, ISSN 0003-2670.

McDowell, W.H. & Likens, G.E. (1988). Origin, composition and flux of dissolved organic carbon in the Hubbard Brook Valley. *Ecological Monographs*, Vol. 58, pp. 177-195, ISSN 0012-9615.

McKnight, D.M., Boyer, E.W., Doran, P., Westerhoff, P.K., Kulbe, T. & Andersen, D.T. (2001). Spectrofluorometric characterization of dissolved organic matter for indication of precursor organic material and aromaticity. *Limnology and Oceanography*, Vol. 46, No. 1, pp. 38-48, ISSN 1939-5590.

Mopper, K. & Schultz, C.A. (1993). Fluorescence as a possible tool for studying the nature and water column distribution of DOC components. *Marine Chemistry*, Vol. 41, No. 1-3, pp. 229-238, ISSN 0304-4203.

Ohno, T. (2002). Fluorescence inner-filtering correction for determining the humification index of dissolved organic matter. *Environmental Science and Technology*, Vol. 36, No. 4, pp. 742-746, ISSN 1520-5851.

Osburn, C.L., Morris, D.P., Thorn, K.A. & Moeller, R.E. (2001). Chemical and optical changes in freshwater dissolved organic matter exposed to solar radiation. *Biogeochemistry*, Vol. 54, pp. 251-278, ISSN 1573-515X.

Pagano, T.E. & Kenny, J.E. (1999). Assessment of Inner Filter Effects in Fluorescence Spectroscopy using the Dual- Pathlength Method– a Study of Jet Fuel JP-4. *Proceedings of SPIE*, Vol. 3856, pp. 289-297.

Patel-Sorrentino, N., Mounier, S., Lucas, Y. & Benaim, J.Y. (2004). Effects of UV-visible irradiation on natural organic matter from the Amazon basin. *Science of the Total Environment*, Vol. 321, No. 1–3, pp. 231-239, ISSN 0048-9697.

Patra, D. & Mishra, A.K. (2002a). Total synchronous fluorescence scan spectra of petroleum products. Analytical and Bioanalytical Chemistry, Vol. 373, No. 4-5, pp. 304-309, ISSN 1618-2650.

Patra, D. & Mishra, A.K. (2002b). Recent developments in multicomponent synchronous fluorescence scan analysis. *Trends in Analytical Chemistry*, Vol. 21, No. 12, pp. 787 – 798, ISSN 0165-9936.

Parker, C.A. & Barnes, W.J. (1957). Some experiments with spectrofluorimeters and filter fluorimeters. *The Analyst*, Vol. 82, No. 978, pp. 606–618.

Parlanti, E., Wörz, K., Geoffroy, L. & Lamotte, M. (2000). Dissolved organic matter fluorescence spectroscopy as a tool to estimate biological activity in a coastal zone submitted to anthropogenic inputs. *Organic Geochemistry*, Vol. 31, No. 12, pp. 1765-1781, ISSN 0146-6380.

Pascu, M.L, Cristu, D., Vasile, A. & Moise, N. (1995). Pesticide monitoring in water using laser-induced fluorescence and absorption techniques. *Proceedings of SPIE*, Vol. 2461, pp. 648-652.

Pharr, D.Y., McKenzie, J.K. & Hickman, A.B. (1992). Fingerprinting Petroleum Contamination Using Synchronous Scanning Fluorescence Spectroscopy. *Groundwater*, Vol. 30, No. 4, pp. 484–489, ISSN 1745-6584.

Pfeiffer, E., Pavelescu, G., Baker, A., Roman, C., Ioja, C. & Savastru, D. (2008). Pollution analysis on the Arges River using fluorescence spectroscopy. *Journal of Optoelectronics and Advance Materials*, Vol. 10, No. 6, pp. 1489 – 1494, ISSN 1841 - 7132.

Reynolds, D. & Ahmad, S.R. (1995). Effect of metal ions on the fluorescence of sewage wastewater. *Water Research*, Vol. 29, No. 9, pp. 2214-2216, ISSN 0043-1354.

Reynolds, D.M. & Ahmad, S.R. (1997). Rapid and direct determination of wastewater BOD values using a fluorescence technique. *Water Research*, Vol. 31, No. 8, pp. 2012-2018, ISSN 0043-1354.

Reynolds, D.M. (2003). Rapid and direct determination of tryptophan in water using synchronous fluorescence spectroscopy. *Water Research*, Vol. 37, No. 13, pp. 3055-3060, ISSN 0043-1354.

Ryder, A. (2005). Analysis of Crude Petroleum Oils Using Fluorescence Spectroscopy, In: *Reviews in Fluorescence*, Eds. Geddes, C.D. , Lakowicz, J.R., Springer US, pp. 169-198.

Selli, E., Zaccaria, C., Sena, F., Tomasi, G. & Bidoglio, G. (2004). Application of multi-waymodels to the time-resolved fluorescence of polycyclic aromatic hydrocarbons mixtures in water. *Water Research*, Vol. 38, No. 9, pp. 2269–2276, ISSN 0043-1354.

Senesi, N. (1993). Nature of interactions between organic chemicals and dissolved humic substances and the influence of environmental factors, In: *Organic substances in soil and water: Natural constituents and their influences on contaminant behaviour*, Eds. Beck, A.J., Jones, K.C., Hayes ,M.H.B., Minglegrin, U., Publisher Royal Society of Chemistry, Cambridge.

Smart, P.L., Finlayson, B.L., Rylands, B.L. & Ball, C.M. (1976). The relation of fluorescence to dissolved organic carbon in surface waters. *Water Research*, Vol. 10, No. 9, pp. 805-811, ISSN 0043-1354.

Spencer, R.G.M., Pellerin, B.A., Bergamaschi, B.A., Downing, B.D., Kraus, T.E.C., Smart, D.R., Dahlgren, R.A. & Hernes, P.J. (2007). Diurnal variability in riverine dissolved organic matter composition determined by in situ optical measurement in the San Joaquin River (California, USA). *Hydrological Processes*, Vol. 21, No. 23, pp. 3181-3189, ISSN 1099-1085.

Spitzy, A. & Leenheer, J.A. (1991). Dissolved organic carbon in rivers, In: *Biogeochemistry of major world rivers*, Eds. Degens, E.T., Kempe, S., Richey, J.E., Publisher Wiley, New York, pp. 213-232.

Stedmon, C.A., Markager, S. & Bro, R. (1993). Tracing dissolved organic matter in aquatic environments using a new approach to fluorescence spectroscopy. *Marine Chemistry*, Vol. 82, No. 3-4, pp. 239-254, ISSN 0304-4203.

Stewart, A.J. & Wetzel, R.G. (1980). Asymmetrical relationships between absorbance, fluorescence, and dissolved organic carbon. *Limnology and Oceanography*, Vol. 26, No. 3, pp. 590-597, ISSN 1939-5590.

Swietlik, J. & Sikorska, E. (2004). Application of fluorescence spectroscopy in the studies of natural organic matter fractions reactivity with chlorine dioxide and ozone. *Water Research*, Vol. 38, No. 17, pp. 3791-3799, ISSN 0043-1354.

Thomas, J.D. (1997). The role of dissolved organic matter, particularly free amino acids and humic substances, in freshwater ecosystems. *Freshwater Biology*, Vol. 38, No. 1, pp. 1-25, ISSN 1365-2427.

Thurman, E.M. (1985). Organic geochemistry of natural waters, In: *Developments in biogeochemistry*, Eds. Niyhoff, M., Junk W., Publisher Springer Science & Business, Dordrecht, The Netherlands.

Valeur, B. (2001). Molecular Fluorescence: Principles and Applications, Publisher Wiley, Weinheim.

4

Grain Size and Source Apportionment of Heavy Metals in Urban Stream Sediments

K. Sekabira[1], H. Oryem-Origa[2], T. A. Basamba[3],
G. Mutumba[4] and E. Kakudidi[5]
*[1]Department of Environment, School of Engineering and Applied Sciences
Kampala International University, Kampala
[2]Department of Biology, School of Biological Sciences, Makerere University, Kampala
[3]Soil Science Department of Agricultural Production
College of Agricultural and Environmental Sciences, Makerere University, Kampala
Uganda*

1. Introduction

Heavy metals are ubiquitous environmental pollutants and concern over possible health risks and ecosystem effects in sediments have increased in recent years. In aquatic sediments, heavy metals are mostly enriched in the fine grained fractions (Salomon and Förstner 1984; Muwanga, 1997; Prego *et al.*, 1999). Studies have therefore used clay (< 2 μm) (Förstner 1987) and < 63 μm (Muwanga 1997) fractions in assessment of heavy metal concentrations in sediments. The specific surface area of sediments is dependant on granulometric parameter and mineral composition (Juracic *et al.*, 1982). Association of metals with smaller grain-size particles is attributed to co-precipitation and complexation of metals on particle surfaces and this determines the distribution pattern of heavy metals in sediments (Ho *et al.*, 2010). Most of the metal content occurs in complex form with insoluble inorganic and organic ligands. Heavy metal emissions from anthropogenic activities occur in stream or river sediments, where they are absorbed onto clays and other fine grained materials (*Ho et al.*, 2010) Heavy metals can be absorbed on negatively charged surfaces of clay minerals, organic matter or iron and manganese oxides and hydroxides. Sediments are considered to be important carriers as well as sinks for heavy metals in the hydrological cycle (Muwanga, 1997). The objectives of this study were: (1) to determine heavy metal content in urban stream sediment fractions and, (2) to assess source apportionment of heavy metals in stream sediments. This study was conducted between the months of August 2008 to November 2009, along the Nakivubo Channelized stream, Kampala Uganda.

2. Materials and methods

2.1 Study area and sampling site

This study was conducted along the Nakivubo Channelized stream in metropolitan Kampala (0°15'N and 32°30'E). Nakivubo channel drains through Kampala city centre and the Upper and Lower Nakivubo swamps before discharging into the Inner Murchison Bay of Lake Victoria. The study area is located 45km north of the equator and 8km north of Lake

Victoria , with a total area of 190 km^2 (Fig. 1). The study area has a tropical climate that is attributed to high altitude, relief, proximity to Lake Victoria and long distance from the sea (Matagi *et al.*, 1998). The Lake Victoria basin has warm temperatures ranging between 23°C to 32°C and a bi-modal rainfall pattern averaging approximately 1260 mm annually. The area is underlain by granites and granitoid gneisses of the Precambrian. These are overlain by phyllites and schist of the Buganda-Toro system with a mixture of alluvial and lacastrine sand, silt and clay that characterise Nakivubo swamp soils. Soils have been derived from weathering of the rocks. The alluvial soils from the upper (profile A) layers are composed of semi-liquid organic material and those underneath consists of reddish ferruginous loams and clays (profile B) attributed to organic decomposition and runoff (Kansiime and Nalubega, 1999).

In this study, the stream was subdivided into three sections (Fig. 1; Table 1) namely; Upstream (US01 – MD05) characterised by commercial establishments, Midstream (MD05 – DS15) characterised by commercial and industrial establishment and Downstream (DS15 and beyond), characterised by the Nakivubo wetland.

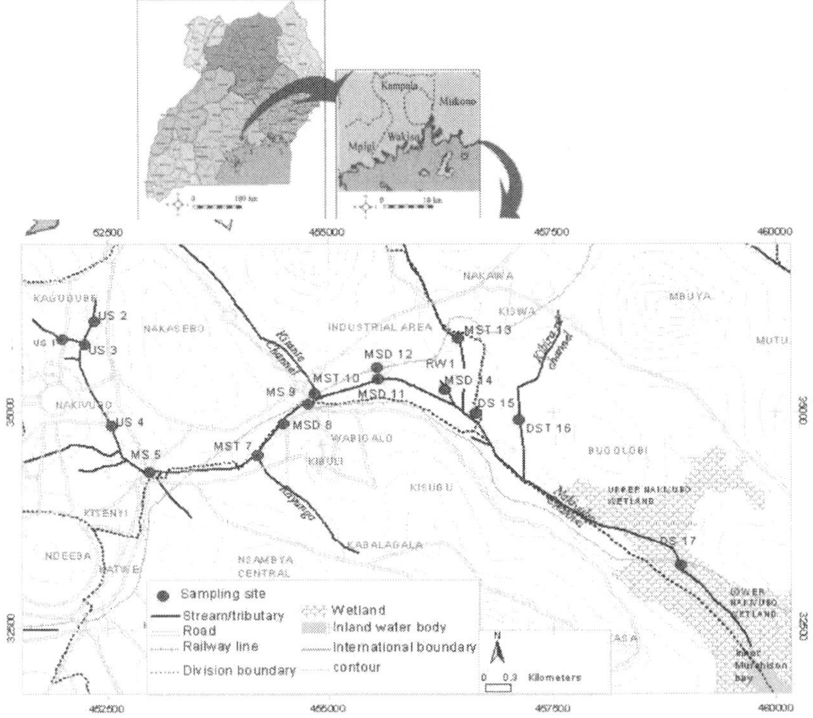

Fig. 1. Map of Kampala showing the locations of the sampling sites along the Nakivubo Channelized stream in Kampala. US-upstream; MS-midstream; MST-midstream tributary; RW-rain water; MSD-midstream discharge point; DS-downstream; DST-downstream tributary

Site/Location	Code	Activity/Establishment	Sediment
Upstream			
Agakhan High School Bridge	US1	Car washing bay, fish factory, gas/fuel station, residential, bus parking yard, seepage from walls	Sand
Bativa Hotel Bridge	US2	Car washing bay, gas/fuel station, slum residential and commercial and seepage	Sand
Kiseka market Bridge	US3	Car washing bay, garage, commercial and seepage	Sand
Nakivubo Stadium Bridge	US4	Recreational, commercial, market, vehicle traffic, bus park, gas/petro station, and seepage	Silty sand
Midstream			
Fire Brigade Bridge	MS5	Commercial, recreational, vehicle traffic, bus park, gas/petro station, cement stores, Katwe metal works and fabrications and seepage	Silty sand
6th Street Bridge Mukwano	MS9	Commercial, oil storage in vicinity, vehicle traffic, gas/petro station, seepage, industries	Silty sand
Downstream			
5th Street Bridge	DS15	Industries, vehicle traffic, sewerage plant, seepage, garages, metal fabrication, petro station , residential	Sand
Luzira Culvert	DS17	Industries, cultivation, fishing, residential	Muddy sand
Tributaries			
Kayunga Stream	MT7	Solid waste dump sites, horticulture, recreational, slum and residential, vehicle traffic, gas/petro station	Sand
Kitante Stream	MT10	horticulture, recreational, residential and commercial, vehicle traffic, gas/petro station	Sand
Lugogo Stream	MT13	Vehicle traffic, commercial, residential and Industrial, electric station, horticulture, carpentry works, pole treatment and seepage	Sand
Kibira Road Stream	DT16	Battery , plastic and paper factory, Industries, and gas/petro station	Sand

US-upstream; MS-midstream; DS-downstream; MT-midstream tributary; DT-downstream tributary

Table 1. Sample site Location, description of activities and sediments type

2.2 Determination of sediment particle size (texture)

Sediment samples were collected between August, 2008 to November, 2009 along the Watindo stream (the control) and Nakivubo Channelized stream and its tributaries (Fig.1), using a hand trowel. Watindo stream was chosen to be outside the study area, 30 km along Gulu road for comparison purposes. The samples were placed in Ziploc bags and transported to the laboratory. The hand trowel [plastic] was washed with a detergent, rinsed and dried before each use so as to minimize contamination. Sediment samples were air dried and 300.0 g were transferred into a set of standard sieves (0.063 mm, 0.125 mm, 0.25 mm, 0.5 mm, 1.0 mm and 2.0 mm) with the largest mesh size on top and the smallest at the bottom. The sieving was carried out using a vibrating machine. The weight of soil retained on each sieve was determined and a cumulative percentage calculated. Coefficient of curvature and uniformity were assessed to determine particle-size distribution fractions of each sample.

The total content of Pb, Cd, Cu, Zn, Mn and Mn in sediment fractions was determined using the method as described by Sekabira *et al.* (2010). Ideally, 1.25 g of each sample was digested with 20 mL aqua regia (HCl/HNO$_3$ 3:1) in a beaker (open-beaker digestion) on a thermostatically controlled hot plate. The digest was heated to near dryness and cooled to ambient temperature. Then 5.0 mL of hydrogen peroxide was added in parts to complete the digestion and the resulting mixture heated again to near dryness in a fume cupboard. The beaker wall was washed with 10 ml of de-ionised water and 5.0 ml HCl were added, mixed and heated again. The resulting digest was allowed to cool and transferred into a 50 mL standard flask and made up to the mark with de-ionised water. Pb, Cd, Cu, Zn, Mn and Fe were then analyzed by direct aspiration of the sample solution into a Perkin-Elmer model 2380 Flame Atomic Absorption Spectrophotometer (AAS). All metals were analysed using lean-blue acetylene flame at wavelength 324.8 nm, slit width 0.2 mm and sensitivity check of 5.0 mg/L Cu; wavelength 228.8 nm, slit width 0.7 and sensitivity check of 2.0 mg/L Cd; wavelength 213.9 nm, slit size 0.7 nm and sensitivity check 9.0 mg/L Pb and wavelength 279.5 nm, slit size 0.2 nm and sensitivity check of 2.5 mg/L Mn. Sediment pH was measured in a suspension of 1:2.5, sediment to water ratio using a calibrated pH meter (WE-30200). Accuracy of the analytical method was evaluated by comparing the expected metal concentrations in certified reference materials with the measured values. Simultaneous performance of analytical blanks, standard reference (JG-3) (Imai *et al.*, 1995) and calculation of the average recoveries of heavy metals confirmed that the accuracy of the method was within acceptable limits (Table 2).

Heavy metals	Pb	Cd	Cu	Zn	Mn (%)	Fe (%)
Reference material	11.7	0.054	6.81	46.5	0.055	2.58
Measured values	10±0.981	0.05±0.002	6.75±0.131	48.25±1.041	0.048±0.003	2.35±0.139
% Recovery	85.5	92.6	99.1	103.8	87.3	91.1

Table 2. Quality control (mean ± SD) (mg/kg trace and % for elements)

2.3 Assessment of heavy metal distribution in sediment

Enrichment Factor (EF): As proposed by Simex and Helz (1981), EF was employed to assess the degree of contamination and to understand the distribution of heavy metal elements of anthropogenic origin from sites by individual elements in sediments. Iron (Fe) was chosen as the normalizing element while determining EF-values, since in wetlands it is mainly supplied from sediments and is one of the widely used reference element (Loska *et al.*, 2002; Kothai et al., 2009: Chakravarty and Patgiri, 2009; Seshan *et al.*, 2010). Other widely used reference metal elements include Al and Mn (Nyangababo *et al.*, 2005a; Kamaruzzaman *et al.*, 2008; Ong and Kamaruzzaman, 2009).

$$\text{Enrichment Factor (EF)} = (C_n/Fe)\ \text{sample}/\ (C_n/Fe)\ \text{background}$$

where C_n is the concentration of element "n". The background value is that of average shale (Turekian and Wedepohl, 1961). Elements which are naturally derived have an EF value of nearly unity, while elements of anthropogenic origin have EF values of several orders of magnitude. Six categories are recognised: ≤ 1 background concentration, 1 - 2 depletion to minimal enrichment, 2 - 5 moderate enrichment, 5 - 20 significant enrichment, 20 - 40 very high enrichment and > 40 extremely high enrichment (Sutherland, 2000).

Analysis of variance (ANOVA): ANOVA was employed to determine whether groups of variables have the same means on data that are continuous or normally distributed and with homogeneous variance. Additionally, it was employed to assess the relationship between heavy metal concentrations and their interaction between sections of the stream.

Correlation analysis: Pearson's correlation analysis was adopted to analyse and establish inter-metal relationship and physico-chemical characteristics of the stream water.

Cluster Analysis (CA) and Factor analysis (FA): CA was performed to classify elements of different sources on the basis of their similarities using dendrograms and to identify relatively homogeneous groups of variables with similar properties. FA was employed on the variables that are correlated to isolate or determine specific factors that are associated with such groupings of metal concentrations so as to establish their origin and distribution. The data was standardised to give a normal distribution with a mean of 0 and a variance of 1. Sample means were standardised by subtracting the mean of their distribution and dividing by standard error (SE) or square root of the variance.

3. Results

3.1 Sediment grain size

Textural composition of the sediment samples is shown in Fig 3. Stream sediments along Nakivubo Channelized stream ranged from 3.0 to 14.0 % clay and silt, 5.0 to 20.0 % fine sand, 15.0 to 29.0 % medium sand and 16.0 to 48.0 % coarse sand grain size fractions. The Nakivubo tributaries ranged from 3.3 to 7.2 % clay and silt, 6.0 to 18.0 % fine sand, 25.0 % to 60.0 % medium sand and 27.0 to 52.0 % coarse sand grain size fractions. Industrial outfall sediments indicated a range of 3.2 to 16.0 % clay-silt, 3.0 to 15.0 % fine sand, 15.0 to 30.0 % medium sand and 31.0 to 53.0 % coarse sand grain size fractions. Watindo stream sediments range from 11 to 16.4 % clay and silt, 14 to 18.0 % fine sand, 17 to 23.0 % medium sand and 23 to 29.0 % coarse sand grain size fractions. However, sediments sampled along the Nakivubo channelized stream can generally be described as coarse grained. The percentage of all fractions showed a similar trend (Fig. 3) along the Nakivubo stream.

3.2 Heavy metal concentrations in various fractions

The mean pH ranged between 5.71±1 (slightly acidic) and 7.25±1 (neutral), but at Watindo stream, the mean pH of the sediments ranged from highly acidic (4.53±1) to acidic (5.40±1) (Table 3). Total heavy metal concentrations in < 63 μm and 63-125 μm sediments as well as distribution pattern along the Nakivubo Channelized stream are indicated in Table 3, Fig. 2 and Fig. 4. The Nakivubo stream sediments showed the highest heavy metal content of Pb (218.64 mg/kg) at Kisekka market in fine sand fractions, Cd (2.46 mg/kg) and Cu (435.96 mg/kg) at Agakhan High School Bridge in fine sand fractions and Zn (261.2 mg/kg) at Luzira culvert in clay-silt fractions. Industrial outfall sludge and sediment samples showed high heavy metal concentration of Pb (132.0 mg/kg), Cu (495.2 mg/kg) and Zn (1361.2 mg/kg) at National Water and sewerage corporation plant in clay-silt fractions and Cd

Sites*	pH-s	Code	Heavy metal (mg/kg) in Clay and silt (< 63 m)						Heavy metal (mg/kg) in very fine sand (63 - < 1255m)					
			Pb	Cd	Cu	Zn	Mn	Fe	Pb	Cd	Cu	Zn	Mn	Fe
Nakivubo Channel														
Agakhan High Bridge	6.99	US01	72.00	1.20	100.40	285.20	880.00	92000.00	73.89	2.46	435.96	197.04	2955.67	295566.50
Bativa Hotel Bridge	7.13	US02	64.00	1.20	75.20	165.60	1000.00	128000.00	35.77	1.43	58.66	123.75	1144.49	207439.20
Kiseka Market Bridge	7.25	US03	144.00	0.80	196.80	353.20	920.00	68000.00	218.64	0.58	86.31	78.25	2819.33	207134.64
Stadium Bridge	6.75	US04	144.00	1.20	117.20	131.60	520.00	52000.00	76.00	0.80	25.20	52.00	80.00	9200.00
Fire Brigade Bridge	7.11	MS05	156.00	1.20	156.00	329.20	840.00	92000.00	112.18	1.60	285.26	189.10	5769.23	560897.44
6th Street Bridge	6.93	MS09	104.00	0.80	96.00	325.20	520.00	56000.00	52.00	0.40	53.20	131.20	320.00	37600.00
5th Street Bridge	6.91	DS15	96.00	1.60	143.20	361.20	1120.00	132000.00	123.20	1.03	103.70	283.37	308.01	31827.52
Luzira Culvert	5.71	DS17	132.00	2.00	190.00	381.20	1520.00	80000.00	80.00	1.60	55.60	261.20	1040.00	60000.00
Industrial outfall														
Mukwano Industries	6.82	MD08	68.00	1.20	103.20	144.40	1560.00	120000.00	40.98	2.05	92.21	153.69	16803.28	1270491.80
Peacock Paint Factory	6.92	MD11	104.00	4.00	351.20	1077.20	400.00	72000.00	93.58	6.68	343.58	1140.37	267.38	33422.46
City Abattoir	7.08	MD12	104.00	1.20	159.60	148.80	880.00	140000.00	82.83	0.75	42.92	126.51	1430.72	195783.13
NWSC Sediments	6.84	MD14	48.00	1.20	64.00	192.00	-	-	64.00	1.20	58.00	186.40	-	-
NWSC Sludge		MD14	132.00	1.60	495.20	1293.20	1360.00	144000.00	124.00	2.00	192.00	1361.20	1520.00	172000.00
Tributaries														
Kitante Stream	6.94	MT10	44.00	0.80	68.40	96.80	680.00	37600.00	28.00	0.80	37.20	68.40	520.00	28400.00
Lugogo Stream	6.55	MT13	84.00	0.80	283.20	577.20	2000.00	216000.00	72.00	0.80	34.80	105.20	1480.00	156000.00
Kibira Rd Stream	6.89	DT16	352.00	0.80	114.40	461.20	480.00	28800.00	192.00	0.80	26.80	146.00	120.00	6000.00
Watindo Stream														
	5.09	CTL1	44.00	0.80	195.60	93.20	1960.00	92000.00	54.05	1.35	86.49	62.16	8513.51	567567.57
	4.53	CTL2	32.00	1.20	97.60	84.00	400.00	60000.00	20.00	0.40	32.40	27.60	120.00	40000.00
	5.4	CTL3	20.00	2.00	47.60	38.00	360.00	64000.00	28.00	1.20	21.60	28.00	280.00	76000.00

Table 3. Total heavy metal content (mg/kg) in the stream sediments (silt-clay < 63 and 63- 125μm) fractions of Nakivubo Channel, its tributaries, industrial discharge outfall and Watindo stream

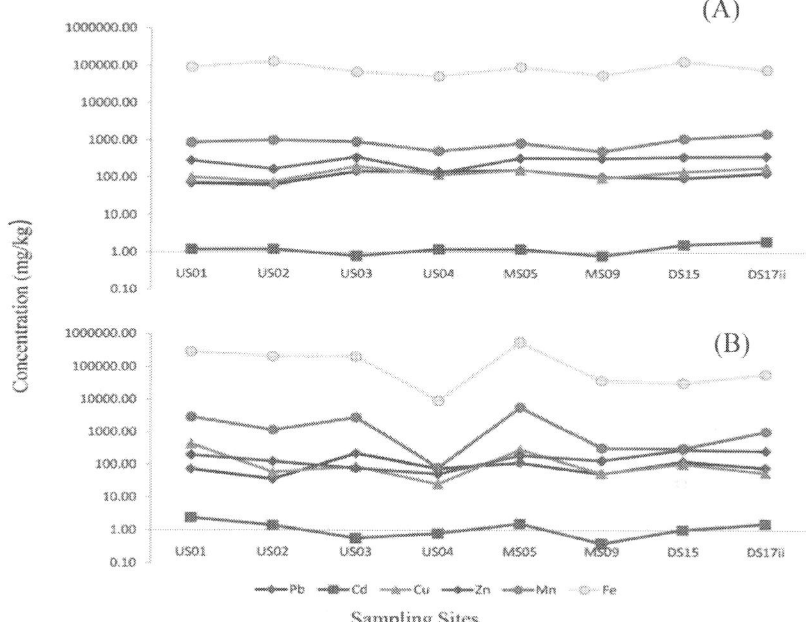

Fig. 2. Heavy metal concentrations along Nakivubo channelized stream in sediment fractions: (A) clay-silt fractions; (B) fine-sand fractions

(4.0 mg/kg) at Peacock paint factory in clay and silt fractions respectively. Nakivubo tributaries also indicated high heavy metal concentration of Pb (352.0 mg/kg) along Kibira Road stream in clay-silt fractions and Cu (283.2 mg/kg) and Zn (577.2 mg/kg) along Lugogo stream in clay-silt fractions. Watindo stream sediments showed high concentration of Pb (54.05 mg/kg) at site CTL1 in fine sand fractions, Cd (2.0 mg/kg) at CTL3 in clay-silt fractions and Cu (195.6 mg/kg) and Zn (93.2 mg/kg) at CTL 1 in clay-silt fractions. Heavy metal concentrations corresponded with the percentage clay-silt fractions in sediments along the Nakivubo channelized stream.

Analysis of variance (ANOVA) was used to determine whether heavy metal variables in the sediment fractions have the same mean on data that are normally distributed. ANOVA results are shown in Table 4. Heavy metal concentrations in sediments along Nakivubo Channelized stream showed significant variation in the means of clay-silt and fine sand fractions for Zn ($F_{1,14}$ = 6.646, $p < 0.05$) whereas, Pb ($F_{1,14}$ = 1.258, $p = 0.281$), Cd ($F1_{,14}$ =0.069, $p = 0.797$) and Cu ($F_{1,14}$ = 0.901, $p = 0.359$) were not significantly different. Mean values for Cu ($F_{1,4}$ = 10.52, $p < 0.05$) and Cd ($F_{1,4}$ = 65535, $p < 0.05$)along Nakivubo tributaries were significantly different whereas, Pb ($F_{1,4}$ = 0.238, $p = 0.651$), and Zn ($F_{1,4}$ = 3.13, $p = 0.152$) showed no significant difference. Lead, cadmium, copper and zinc showed no significant ($p > 0.05$) difference in the means of the elemental concentrations along Watindo stream (Table 4). However, the mean values were higher for clay-silt fractions than fine sand.

ANOVA showed no significant variation in the means of clay-silt and fine sand in Pb, Cd and Cu elements Mean concentrations of clay-silt fractions were higher than the mean of fine sand fractions. Clay-silt and fine sand sediment fractions accumulated Pb, Cd, Cu, Zn, Mn and Fe. Concentrations of Cu in tributaries, Zn along the Nakivubo Channelized stream and Mn and Fe elements were significantly high in clay-silt fractions (< 63 μm). EF values of heavy metals in the clay-silt fractions showed a relatively homogeneous distribution pattern within the Upstream and Midstream section, suggesting local pollution and terrigenous influences.

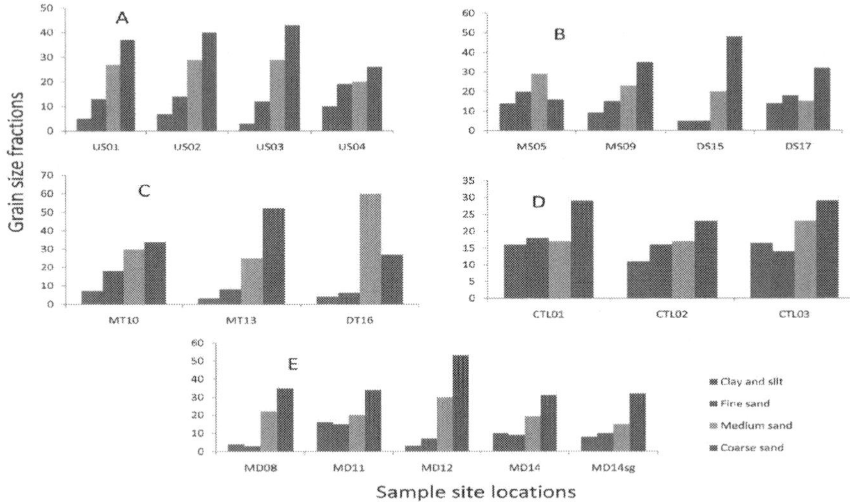

(A and B); tributaries (C); Watindo (D); industrial outfall (E)

Fig. 3. Textural composed of the sediment grain-size fractions in Nakivubo stream sediments

Source of Variation	Dependent variables	SS	DF	MS	F	p
Nakivubo Channel sites	Pb	0.050	1	0.050	1.258	0.281
	Cd	0.001	1	0.001	0.069	0.797
	Cu	0.085	1	0.085	0.901	0.359
	Zn	0.316	1	0.316	6.646	**0.022**
Nakivubo tributaries Site	Pb	0.046	1	0.046	0.238	0.651
	Cd	0.000	1	0.000	65535	**0.051**
	Cu	0.543	1	0.543	10.520	**0.032**
	Zn	0.322	1	0.322	3.130	0.152
Industrial outfall sites	Pb	0.007	1	0.007	0.223	0.649
	Cd	0.009	1	0.009	0.204	0.664
	Cu	0.117	1	0.117	0.858	0.381
	Zn	0.000	1	0.000	0.000	0.995
Watindo stream	Pb	0.000	1	0.000	0.004	0.951
	Cd	0.008	1	0.008	0.560	0.496
	Cu	0.231	1	0.231	2.426	0.194
	Zn	0.105	1	0.105	2.426	0.194

Table 4. One-way ANOVA results for sites and mean heavy metal concentration variables (Dependent variables were log-normal transformed)

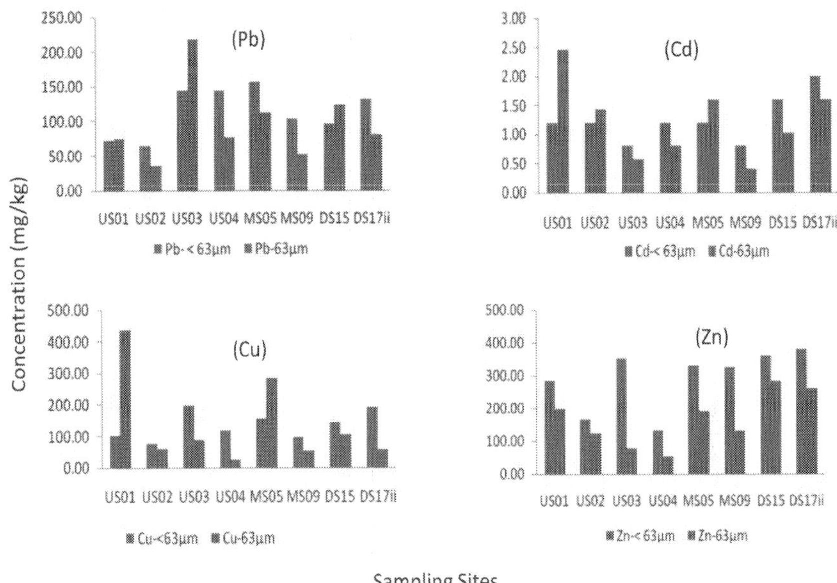

Fig. 4. Heavy metal contents of Pb, Cd, Cu and Zn in the fine fractions (< 63 µm and 63-125 µm) of Nakivubo stream sediments

3.3 Sediment enrichment

The results show that enrichment factor values can be assessed with respect to the average shale in reference to the degree of contamination (Harikumar et al., 2009; Ong and Kamaruzzaman, 2009). Enrichment factor values for fine sediments were highest at Nakivubo Stadium Bridge for Pb (19.0), Cd (13.33), Cu (2.8), Zn (2.74) and Mn (0.47) in fine sand fraction. The sequence of elemental enrichment in sediment fractions followed a decreasing order of Pb > Cd > Cu > Zn > Mn in clay-silt and fine sand fractions at Fire Brigade Bridge, whereas, sediments at 5th Street Bridge showed a decreasing sequence of Pb > Cd > Zn > Cu > Mn in clay-silt and fine-sand fractions. Lead and cadmium in sediments are significantly enriched (5-20), Cu and Zn are moderately enriched (2-5) and Mn was within background concentration (≤1) in fine sand fractions. Generally, enrichment factor in clay-silt fractions for Pb, Cd, Cu, Zn and Mn increased downstream, whereas the EF values for fine sand fractions showed a gradual decrease. Manganese EF values showed background concentrations in clay-silt and fine sand fractions (< 1) (Fig. 5). EF values of heavy metals (Pb, Cd, Cu, Zn and Mn) in clay-silt fractions showed a relatively homogeneous distribution pattern along the Nakivubo stream sediments. The heavy metal concentrations in clay-silt (< 63µm) and fine sand (63- < 125 µm) fractions are within the same order of magnitude, with some variations in concentrations at different sites.

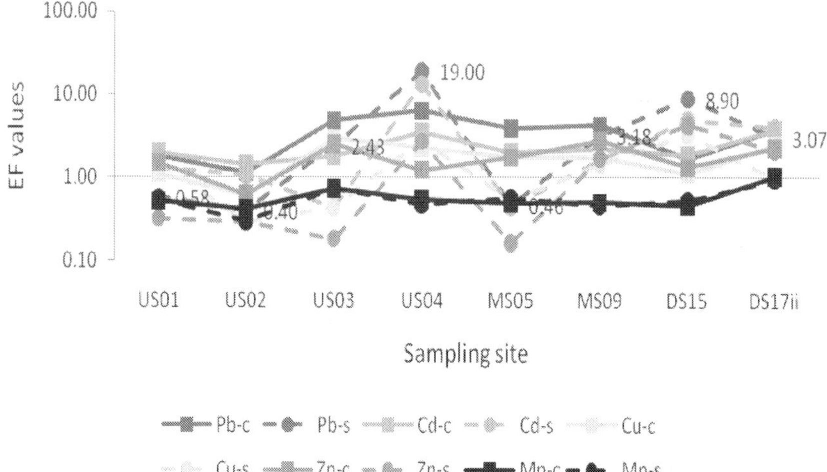

Fig. 5. Distribution of enrichment factor values of Pb, Cd, Cu, Zn, Mn and Fe along the Nakivubo Channelized stream sediment fractions; c-clay-silt fractions; s-fine sand fractions

Results of Pb-Fe, Cd-Fe, Cu-Fe, Zn-Fe and Mn-Fe scatter plots are shown in Fig. 6a. A very poor correlation between naturally occurring concentrations of Fe and other metals (Pb, Cd, Cu and Zn) sampled except Mn which was positively correlated. This may suggest anthropogenic influences of Pb, Cd, Cu and Zn and Mn as a naturally occurring metal concentration in clay-silt fractions. Heavy metals were weakly adsorbed to iron oxides in the clay-silt fractions. At neutral pH, clays have strong negative surface charges that attract Pb, Cd, Cu and Zn cations and iron oxides and hydroxides with positive surface charges.

Results for Pb-Mn, Cd-Mn, Cu-Mn and Zn-Mn scatter plots are shown in Fig. 6b. A linear correlation between Fe-Cd and Fe-Cu elemental pairs suggest that Cd and Cu were naturally occurring heavy metal concentrations (terrigenous) and the outliers would suggest anthropogenic sources. Cadmium and Copper were strongly adsorbed to manganese oxides and hydroxides in the clay-silt fractions. However, Pb and Zn were poorly correlated with Mn suggesting anthropogenic influence.

Results for Pb-Fe, Cd-Fe, Cu-Fe and Zn-Fe scatter plots are shown in Fig.7a. A linear correlation between Fe-Cu and Fe-Mn elemental pairs suggest that Cu and Mn were naturally occurring heavy metal concentrations (terrigenous) and the outliers above the threshold would suggest an anthropogenic source. However, Pb, Cd and Zn were poorly correlated with Mn suggesting anthropogenic influence in fine-sand fractions.

Results for Pb-Mn, Cd-Mn, Cu-Mn and Zn-Mn scatter plots are shown in Fig.7b. A linear correlation between Fe and Cu elemental pairs suggest that Cu was naturally occurring heavy metal concentrations (terrigenous) and the outliers above the threshold would be regarded as anthropogenic. Copper showed strong adsorption to manganese oxides and hydroxides in the clay-silt fractions. However, Pb, Cd and Zn were poorly correlated with Mn suggesting anthropogenic influence in fine-sand fractions.

Heavy metal concentration data in the sediment fractions were subjected to ANOVA and showed no significant variation in the means of clay-silt and fine sand for Pb, Cd and Cu, except for Zn along the Nakivubo Channelized stream (Kruopiene, 2007), Cd and Cu along the tributaries. However, the mean values were higher in clay-silt fractions in all the samples. Elemental concentrations were within the same order of magnitude as observed by Sekabira *et al.*, (2010) along the Nakivubo drainage system except for Cu and Zn which showed extreme high elemental concentrations in sediment fractions of clay-silt and fine sand. Enrichment Factor values of heavy metals in clay-silt fractions (< 63 μm) for Pb, Cd, Cu, Zn and Mn increased gradually downstream, whereas the EF values for fine sand fractions showed an irregular decrease. This phenomenon of increasing heavy metal concentration downstream may be attributed to the increased pollution downstream and adsorption of heavy metals from the water by fine grained sediments with large surface area and clay with negative surface charge. Irregular distribution of heavy metals in fine sand fractions may indicate a localized source of the pollutants, sink and/or retention phenomena (Zanganeh *et al.*, 2008).

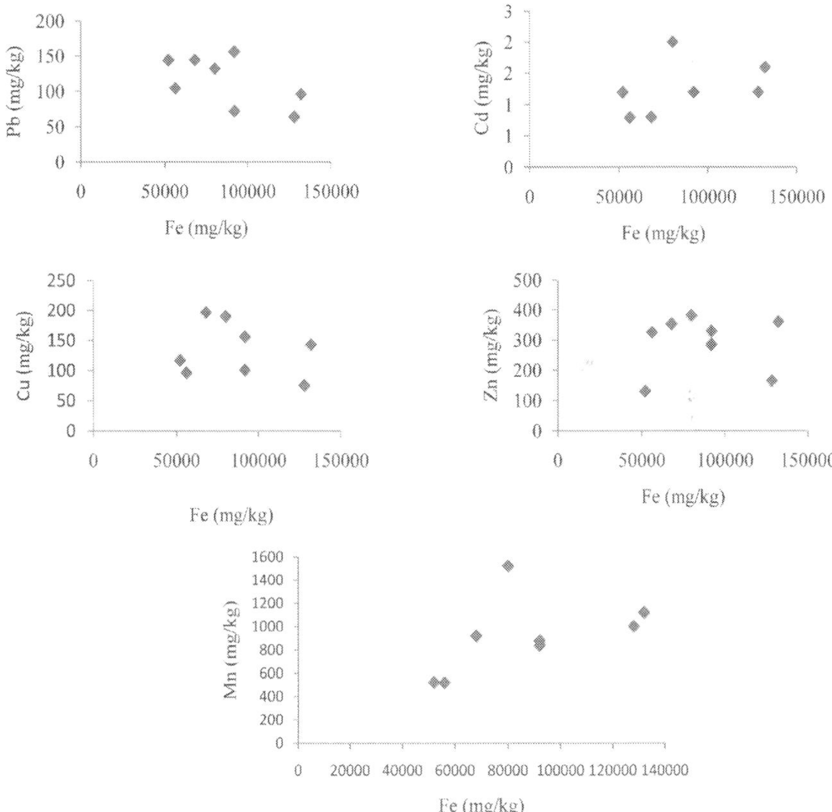

Fig. 6. a. Scatter plots of heavy metals of Nakivubo Channelized stream in clay-silt fractions

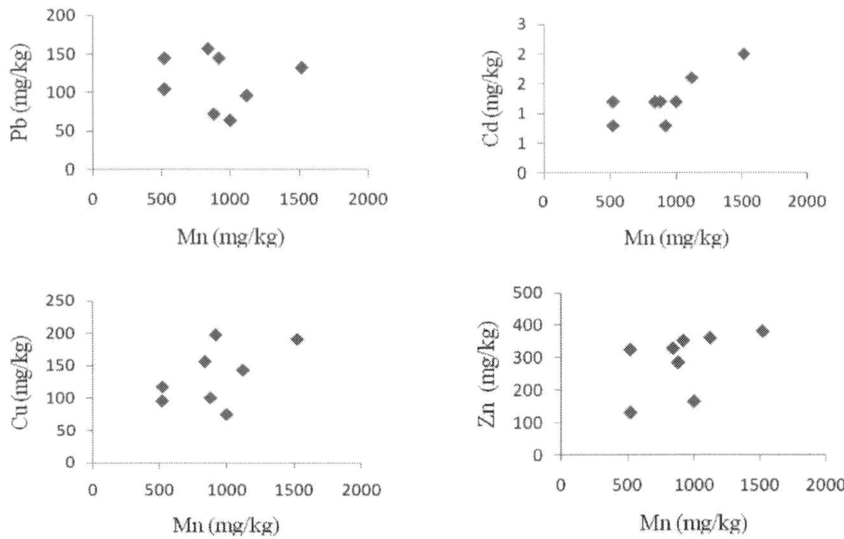

Fig. 6. b. Scatter plots of heavy metals of Nakivubo Channelized stream in clay-silt fractions

3.4 Source apportionment of pollutants

Inter-metal and sediment property association was also evaluated by using Pearson correlation coefficient (r) and the results are presented in Table 5. Results show that elemental pairs Cu-c/Pb-c (r = 0.52 at P = 0.05), Zn-c/Pb-c (r = 0.706, P = 0.01), Pb-s/Pb-c (r = 0.841, P = 0.01), Zn-s/Pb-c (r = 0.487, P = 0.05), Cd-s/Cd-c (r = 0.767, P = 0.01), Zn-s/Cd-c (r = 0.521, P = 0.05), Zn-c/Cu-c (r = 0.773, P = 0.01), Pb-s/Cu-c (r = 0.581, P = 0.01), Cu-s/Cu-c (r = 0.463, p = 0.05), Zn-s/Cu-c (r = 0.646, P = 0.01), Pb-s/Zn-c (r = 0.714, P = 0.01), Cu-s/ Zn-c (r = 0.545, P = 0.05), Zn-s/Zn-c (r = 0.853, P = 0.01), Zn-s/Pb-s (r = 0.528, P = 0.05), Cu-s/Cd-s (r = 0.720, P = 0.01), Zn-s/Cd-s (r = 0.686, P = 0.01), Zn-s/Cu-s (r = 0.693, P = 0.01), Mn-s/Cu-s (r = 0.485, P = 0.05), Fe-s/Cu-s (r = 0.461, P = 0.05), Fe-s/Mn-s (r = 0.943, P = 0.01), BOD/pH-s (r = 0.525, P = 0.05), BOD/Pb-c (r = 0.502, P = 0.05), BOD/Pb-s (r = 0.514, P = 0.05), BOD/Zn-s (r = 0.564, P = 0.05) and Cd-c/% clay-silt (r = 0.481, P = 0.05) are significantly correlated with each other. Lead in clay-silt fractions (Pb-c), Zn-c, Pb-s, Zn-s and BOD were significantly associated with sediment pH, suggesting its influence as a controlling factor. Elemental associations were assessed using Pearson correlation coefficient (r) and indicated that each paired elements had an identical source, geochemistry, and/or common sink (Nyangababo et al., 2005b; Sekabira et al., 2010). Cadmium elemental association with grain size fraction contents may signify its influence as a controlling factor. In aquatic sediments, heavy metals are mostly enriched in and are associated with the fine grained fractions (Muwanga, 1997; Prego et al., 1999; El-Moselhy and Abd El-Azim, 2005). Association of copper in sediments with Mn and Fe-oxides/hydroxides may suggest specific adsorption and co-precipitation by isomorphic substitution.

	pH	Pb-c	Cd-c	Cu-c	Zn-c	Mn-c	Fe-c	Pb-s	Cd-s	Cu-s	Zn-s	Mn-s	Fe-s	% Clay-silt	% Fine-sand
pH-s	1.000														
Pb-c	0.571*	1.000													
Cd-c	-0.068	-0.069	1.000												
Cu-c	0.122	0.502*	0.272	1.000											
Zn-c	0.500*	0.706**	0.269	0.773**	1.000										
Mn-c	0.106	0.050	-0.275	0.352	0.216	1.000									
Fe-c	0.140	-0.149	0.096	0.338	0.231	0.713**	1.000								
Pb-s	0.543*	0.841**	0.020	0.581**	0.714**	0.185	0.051	1.000							
Cd-s	0.182	0.047	0.767**	0.381	0.418	0.044	0.198	0.136	1.000						
Cu-s	0.339	0.202	0.375	0.463*	0.545*	0.208	0.301	0.358	0.720**	1.000					
Zn-s	0.497*	0.487*	0.521*	0.646**	0.853**	0.251	0.335	0.528*	0.686**	0.693**	1.000				
Mn-sa	0.161	-0.072	-0.217	0.192	0.038	0.743**	0.547*	0.088	0.240	0.485*	0.168	1.000			
Fe-s	0.033	-0.270	-0.124	0.116	-0.079	0.631**	0.652**	-0.088	0.231	0.461*	0.082	0.943**	1.000		
% Clay-silt	-0.510	-0.352	0.481*	-0.084	-0.204	-0.271	-0.218	-0.365	0.320	0.095	-0.001	-0.180	-0.096	1.000	
% Fine-sand	-0.306	-0.177	0.095	-0.034	-0.163	-0.312	-0.358	-0.232	0.011	0.044	-0.213	-0.211	-0.167	0.680**	1.000
BOD	0.525*	0.502*	0.264	0.337	0.398	0.083	0.272	0.514*	0.264	0.227	0.564*	0.052	0.001	-0.199	-0.392

**Correlation is significant at the 0.01 level (2-tailed); *. Correlation is significant at the 0.05 level (2-tailed); BOD-biological oxygen demand
s- silt-clay fractions; s- Fine-sand fractions

Table 5. Pearson correlation coefficient (r) matrix of heavy metals, sediment property and BOD in Nakivubo Channelized stream sediments (n=19)

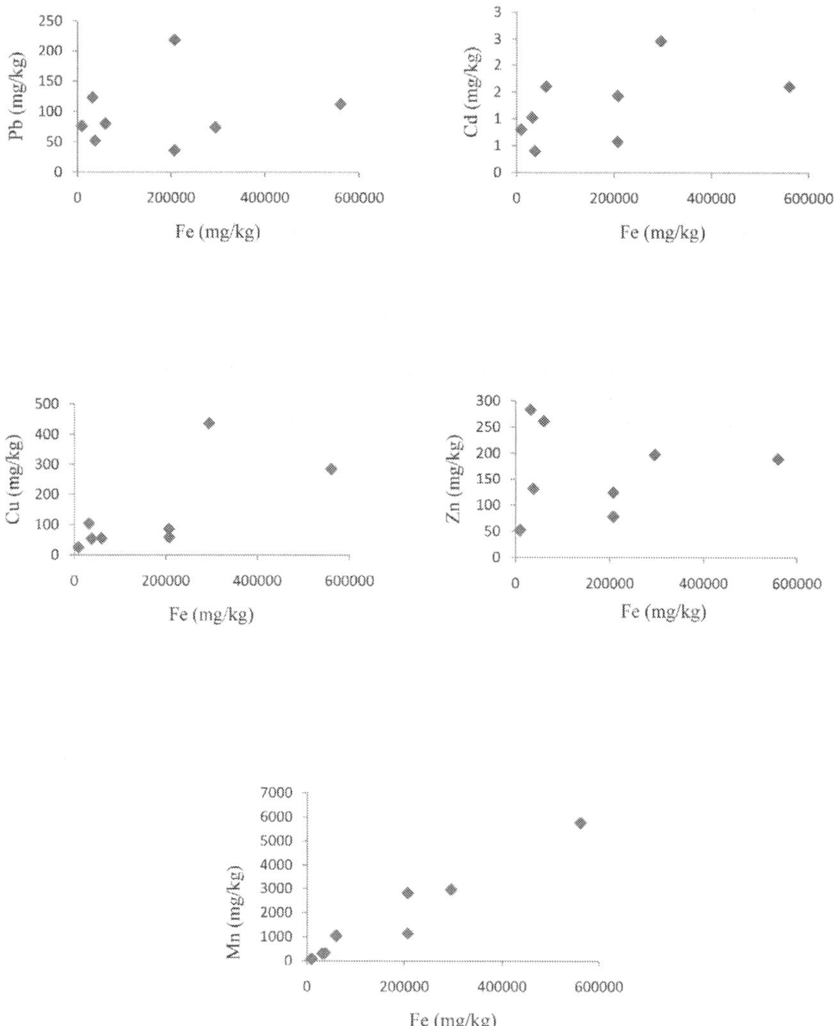

Fig. 7. a. Scatter plots of heavy metals of Nakivubo Channelized stream in fine-sand fractions

Cluster Analysis (CA) was performed on the data using Ward or Average linkage and correlation coefficient distance. Results for CA are shown in Fig. 8. Four clusters of elemental associations were identified based on the fusion of the clusters that are similar. The dendrogram explains the influence and association of the heavy metal clusters or groups by their relative elemental concentrations at each site. CA showed the association of pH and BOD with Pb and Zn in clay-silt and fine-sand fractions as well as Cu in clay-silt fraction in the first group (I). This may suggest the association of Pb with organic matter and pH as a controlling factor. The second group (II) showed the association of Cd and Cu in fine-sand fractions as well as Cd in clay-silt fractions in the Nakivubo stream sediments. Elements in group III (Mn and Fe) originate from terrigenous sources (Sekabira *et al.*, 2010) in both clay-silt and fine-sand fractions. Group IV contains percentage fractions of clay-silt and fine-sand. A biplot of sites and elemental concentrations associated Agakhan High School Bridge and Lugogo stream with Cu and Cd (Fig. 9). This may be attributed to car washing bay, petrol stations and vehicular emissions. Sludge at National Water and Sewerage Corporation contained the highest concentrations of Cu and Zn in both fractions followed by Kiseka Market Bridge attributed to car washing bay and garages. Lead and Zinc concentrations were highest at DT16 site attributed to Uganda batteries limited factory, Uganda Baati limited [galvanised iron sheets] and plastic factory [Uganda house of plastics]. Peacock paint factory is a source of Pb in fine-sand and clay-silt fractions. Cadmium in clay-silt fraction was highest at Nakivubo stadium bridge [US4] attributed to vehicular emissions, car park and a petrol station.

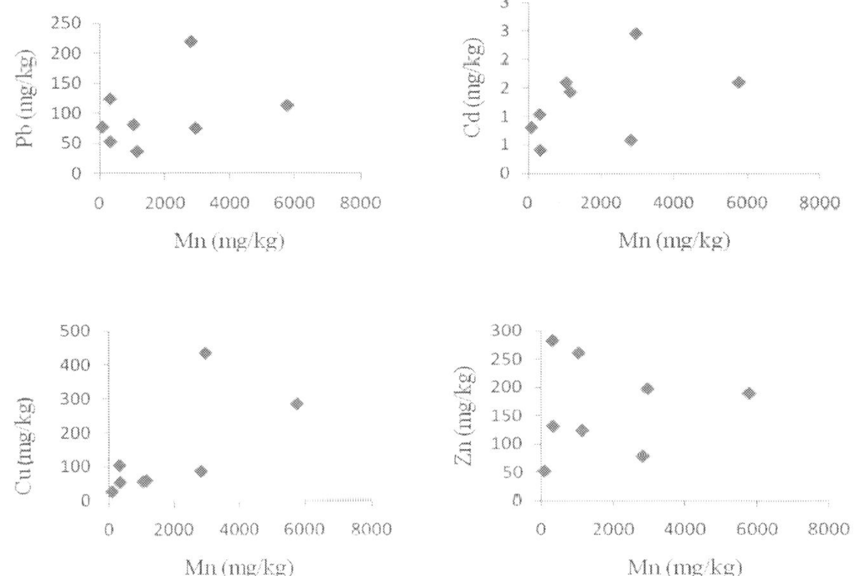

Fig. 7. b. Scatter plots of heavy metals of Nakivubo Channelized stream in fine-sand fractions

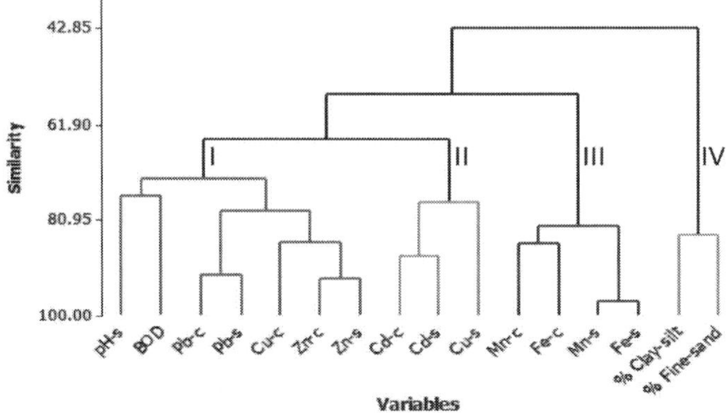

Fig. 8. Dendrogram of urban stream sediment samples along the Nakivubo drainage ecosystem and Watindo stream

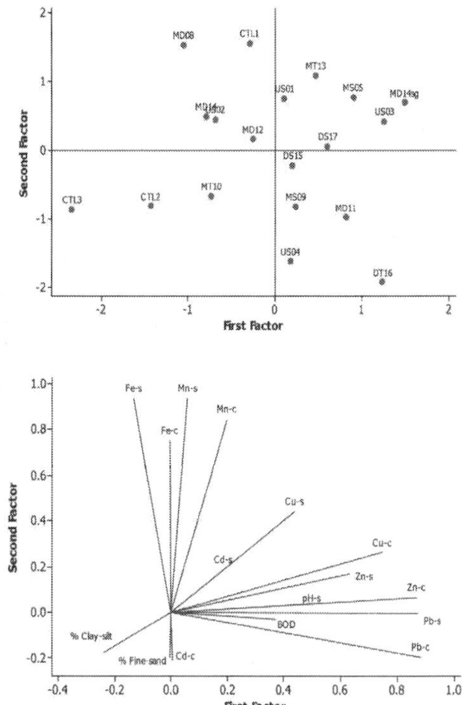

Fig. 9. Biplot of sites and elemental concentrations in Nakivubo drainage system, industrial discharge outfall and Watindo stream; BOD-biological oxygen demand

Variable	Factor 1	Factor 2	Factor 3	Factor 4	Communality
pH-s	**0.502**	0.038	0.045	-0.586	0.598
Pb-c	**0.883**	-0.198	-0.111	-0.249	0.894
Cd-c	-0.004	-0.2	**0.925**	0.029	0.896
Cu-c	**0.748**	0.261	0.273	0.094	0.711
Zn-c	**0.868**	0.064	0.31	-0.143	0.875
Mn-c	0.197	**0.844**	-0.15	-0.112	0.785
Fe-c	-0.005	**0.752**	0.213	-0.31	0.708
Pb-s	**0.872**	-0.004	-0.021	-0.258	0.827
Cd-s	0.182	0.186	**0.891**	0.024	0.862
Cu-s	**0.438**	**0.439**	**0.586**	0.091	0.737
Zn-s	**0.632**	0.166	**0.647**	-0.236	0.901
Mn-s	0.057	**0.935**	-0.004	-0.031	0.877
Fe-s	-0.134	**0.936**	0.082	-0.008	0.901
% Clay-silt	-0.242	-0.176	**0.493**	**0.69**	0.809
% Fine-sand	0.006	-0.213	0.063	**0.859**	0.787
BOD	**0.371**	-0.03	**0.341**	-0.63	0.652
Variance	3.9859	3.5124	3.0339	2.2874	12.8197
% Var.	0.249	0.22	0.19	0.143	0.801

Table 6. Varimax rotated factor loadings and communalities of the Nakivubo stream, tributaries, industrial outfall sediment fractions and Watindo stream (n=15)

Factor Analysis was carried out to establish the influence of sediment grain size, pH and BOD on heavy metal concentrations in clay-silt and fine-sand fractions (Table 6). Four factors with eigenvalues > 1 were extracted in the analysis to help explain the data. The first four factors account for 80.1 % of the total variance/inertia in the data set. The rotated factor matrix is explained by four factors with high communalities of elements except pH. The first factor accounts for 24.9 % of the total variance and contains Pb, Cu and Zn, as well as BOD in water and pH with high variable loading on this factor and corresponds to group I of the cluster analysis. This suggests pH as the controlling factor (Muwanga, 1997; Prego *et al.*, 1999; Abílio *et al.*, 2006 and Ho *et al.*, 2010) and the influence of organic matter on Pb. At neutral pH, clays possess negative surface charges that attract Pb, Cu nd Zn cations into bottom sediments. The second factor accounts for 22.0 % of the total inertia and contains Mn and Fe as well as Cu in fine sand fractions with high variable loadings and corresponds to group III of cluster analysis. This association may be due to their common occurrence in the basic rocks [terrigenous], since their concentrations were within background values (EF ≤ 1) (Sekabira *et al.*, 2010).   The Third factor accounts for 19.0 % of the total inertia and contains Cd, Cu and Zn in fine-sand fractions as well as clay-silt fractions and BOD. The association of Cd and Zn may be attributed to their similar geochemistry and may indicate a source of mixed origin and/or sink of vehicular emissions, Katwe metal works and cement stores at Good Shade. The association of heavy metals and BOD may suggest the role of organic matter (OM) in heavy metal sequestration and the dual origin of Cu and Zn. This causes the transfer of heavy metals into bottom sediments. The Fourth factor accounts for 14.3 % of the total inertia and contains percentage clay-silt and fine-sand fractions with high variable loadings and corresponds to group IV of the cluster analysis.

4. Conclusions

1. Clay-silt fraction sorting increased for zinc concentrations in the Nakivubo stream sediments and generally for lead. Heavy metal concentration increased downstream with percentage increase in clay-silt fractions and enrichment in the very fine grained fractions, probably due to their negative surface charge and higher particulate surface area.
2. The distribution patterns of the heavy metals are controlled by the sorting of fine-grained fractions, pH and organic matter as indicated by BOD.
3. This study showed that stream sediments have background concentrations for Fe and Mn at most of the sites.
4. Factor analysis also indicated three sources of pollutants; (1) mixed origin or retention phenomena of Pb, Cu and Zn as well as BOD of industrial and municipal waste effluents [NWSC]; (2) industrial and vehicular emissions of Cd, Cu and Zn and terrigenous fraction sources characterised by Cu, Mn and Fe.

5. Acknowledgement

The authors are grateful to Kampala International University for the financial support in form of a PhD research project and the Department of Geology, Faculty of Science, Makerere University, for laboratory analyses.

6. References

Abílio, G. Cupelo, A. C. G. and Rezende, C. E. (2006). Heavy metal distribution in sediments of an offshore exploration area, Santos basin, Brazil.

Chakravarty, M. and Patgiri, A. D. (2009). Metal pollution assessment in sediments of the Dikrong River, N.E. India., *J. Hum. Ecol.*, 27 (1), 63-67

El-Moselhy, K. H and Abd El-Azim, H., (2005). Heavy metal content and grain size of sediments from Suez Bay, Red Sea, Egypt. Egyptian Journal of Aquatic Research 31: 2 224-238.

Harikumar, P. S. Nasir ,U. P. and Mujeebu Rahman, M. P. (2009). Distribution of heavy metals in the core sediments of a tropical wetland system. Int., J. Environ. Sci. Tech., 6 (2), 225-232

Ho, H. H. Swennen, R. and Van Damme, A. (2010). Distribution and contamination status of heavy metals in estuarine sediments near Cua Ong Harbor, Ha Long Bay, Viatinam. *Geologica Beligica.*, 13 (1), 37-47.

Imai, N. Terashina, S. Itoh, S. and Ando, A. (1995). Compilation of analytical data for minor and trace elements in seventeen GSJ Geochemical reference samples "Igneous rock series". *Geo-standard Newsletter.*, 19 , 135-213

Juracic, M. Bauman, I. and Pavdic, V. (1982). Are the sediments the ultimate depository of hydrocarbon pollutions?. Ves. J. Etud. Pollut. Mar. Mediterranee CIESM, Cannes, 83-87.

Kamaruzzaman, M. C. Ong, M. S. Noor, A. Shahbudin, S. and Jalal, K. C. A. (2008). Geochemistry of sediments in the major estuarine mangrove forest of Terengganu Region, Malaysia., *American Journal of Applied Science.*, 5 (12), 1707-1712.

Kansiime, F. and Nalubega, M. (1999). Wastewater treatment by a natural wetland: the Nakivubo swamp, Uganda. Processes and implementations. PhD, dissertation, Wageningen Agricultural University, Netherlands.

Kothai, P. Prathibha, P. Saradhi, I. V. Pandit, G. G. and Puranik, V. D. (2009). Characterization of atmospheric particulate matter using pixe technique. *International Journal of Environmental Science and Engineering.*, 1 (1), 27 - 30.

Kruopiene, J., (2007). Distribution of heavy metals in sediments of the Nemunas River (Lithuania). Polish J. of Environ. Stud. 16: 2, 715-722.

Loska, K. Wiechula, D. Barska, B. Cebula, E. and Chojnecka, A. (2002). Assessment of arsenic enrichment of cultivated soils in Southern Poland., *Polish Journal of Environmental Studies.*, 12 (2), 187 – 192.

Matagi, S. V. Swai, D. and Mugabe, R. (1998). A review of heavy metal removal mechanisms in wetlands., Afr. J. Trap. Hydrobiol., Fish 8 , 23-35.

Muwanga, A. (1997). Environmental impacts of copper mining at Kilembe, Uganda: A geochemical investigation of heavy metal pollution of drainage waters, stream, sediments and soils in the Kilembe valley in relation to mine waste disposal. PhD, dissertation. Universitaet Braunschweig, Germany.

Nyangababo, J. T. Henry, I. and Omutunge, E. (2005a). Heavy metal contamination in plants, sediments and air precipitation of Katonga, Simiyu and Nyanduo wetlands of Lake Victoria basin, East Africa. Bull Environ., Contam. Toxicol., 75 (1), 189-196.

Nyangababo, J. T. Henry, E. and Omutange, E. (2005b). Lead, cadmium, copper, manganese and zinc in wetland waters of Victoria lake basin, East Africa., *Bull. Environ. Contam. Toxicol.*, 74 (5), 1003-1010

Ong, M. C. and Kamaruzzaman, B. Y. (2009). An assessment of metal (Pb and Cu) contamination in bottom sediments from South China Sea coastal waters, Malaysia., *American Journal of Applied Science.*, 6 (7), 1418-1423.

Prego, R. Belzunce, M. J. Helios-Rybicka, E. and Barciela, M. C. (1999). Cadmium, manganese, nickel and lead contents in surface sediments of the lower Ulla River and its estuary (Northwest Spain). Bol. Inst. Esp. Oceanogr. 15, (1-4), 495-500.

Salomon, W. and Förstner, U., (1984). Metals in the hydrocycle. Springer-Verlag, New-York Berlin Heidelberg, 349.

Sekabira, K. Oryem-Origa, H. Basamba, T. A. Mutumba, G. and Kakudidi, E. 2010. Assessment of heavy metal pollution in the urban stream sediments and its tributaries. *Int. J. Environ. Sci. Tech.*, 7 (3), 435-446.

Seshan, B. R. R. Natesan, U. and Deepthi, K. 2010. Geochemical and statistical approach for evaluation of heavy metal pollution in core sediments in southeast coast of India., *Int. J. Environ. Sci. Tech.*, 7 (2), 291-306.

Simex, S. A. and Helz, G. R. (1981). Regional geochemistry of trace elements in Chesapeake Bay., Environ. Geol., 3 , 315-323.

Sutherland, R. A. (2000). Bed sediment associated trace metals in an urban stream Oahu, Hawaii., *Environ. Geol.*, 39 (6), 611-627.

Turekian, K. K. and Wedepohl, K. H. (1961). Distribution of the elements in some major units of the earth's crust. *Geol. Soc. American Bull.*, 72 (2), 175-192.

Zanganeh, A. H. P, Lachan, V. C. and Vazyari, M., (2008). Geochemical associations and grain size partitioning of heavy metals in nearshore sedimens along the Iranian Coast of the Caspian Sea. The 12th World Lake Conference: 198-202.

5

Metagenomics in Polluted Aquatic Environments

Alexander M. Cardoso[1,5], Felipe H. Coutinho[1] et al.[*]
[1]Instituto de Bioquímica Médica, Universidade Federal do Rio de Janeiro
[5]Instituto Nacional de Metrologia, Qualidade e Tecnologia
Brazil

1. Introduction

Metagenomics is defined as the culture-independent genomic analysis of biological assemblages providing access to the whole set of genes and genomes from a sample. It encompasses a variety of techniques that are based on (i) total DNA extraction from samples followed by PCR amplification of specific genes, (ii) library construction or amplification and sequencing of the whole genetic material. These methodologies have successfully been applied in studies of composition, dynamics, and functions of microbial communities in a variety of ecosystems including those subjected to anthropogenic modifications (Gilbert & Dupont, 2011).

Culture independent methods allow the analysis of a set of metabolic genes from microbial communities, which can be used to determine how environmental conditions such as pollution can shape community composition and the diversity of genes associated with biogeochemical cycles such as those of carbon, nitrogen, and phosphorus (Singh et al., 2009). This approach is also useful for the discovery of novel environmental microorganisms and genes, with important applications for biotechnology, medicine, and bioremediation (Cardoso et al., 2011).

This applicability has resulted in a recent sharp increase in studies focusing in the metagenomic analysis of polluted sites. Their aim is to characterize microbial communities from a diverse set of environments such as freshwater, marine sediments, open ocean, pelagic ecosystems, soil, and host-associated communities. An example of these initiatives is the Global Ocean Sampling Expedition (GOS), which assessed the genetic diversity of marine microbial communities around the Earth. Since 2003, an enormous amount of data has been generated by GOS helping scientists to reveal the microbial diversity and also allowing them to better understand microbial phylogeny and ecology (Gilbert & Dupont, 2011).

---

[*] Felipe H. Coutinho[1], Cynthia B. Silveira[1], Barbara L. Ignacio[1], Ricardo P. Vieira[1], Gigliola R. Salloto[2], Maysa M. Clementino[2], Rodolpho M. Albano[3], Rodolfo Paranhos[4], Orlando B. Martins[1]
[1]Instituto de Bioquímica Médica, Universidade Federal do Rio de Janeiro
[2]Instituto Nacional de Controle de Qualidade em Saúde
[3]Departamento de Bioquímica, Universidade Estadual do Rio de Janeiro
[4]Instituto de Biologia, Universidade Federal do Rio de Janeiro
[5]Instituto Nacional de Metrologia, Qualidade e Tecnologia
Brazil

2. Metagenomics and bioinformatics

Microorganisms can be found across all environments. Adequate sample collection is the initial and essential step to achieve a comprehensive coverage of the microbial diversity. For aquatic studies, the collection of large volumes of water followed by filtration is recommended since it increases the chance of retrieving rare groups. After sample collection, an optional enrichment culture step can be performed to maximize the abundance of a targeted group of microorganisms by providing its ideal growth conditions. Further screening for specific phenotypic features may be performed during the cultivation step in order to target microbial species of unusual metabolism.

Genetic material can be obtained from samples by several methods that include physical and/or chemical cell lysis followed by extraction and purification of nucleic acids (DNA or RNA). However, the amount of non-biological matter associated with the biological material may interfere with the extraction, quantification and amplification processes. Thus, different samples will require different extraction methods as a good representation of the biological diversity relies on the efficiency of the nucleic acid extraction step. The extracted genetic material must be free of amplification inhibitors and special attention must be given when dealing with samples retrieved from polluted sites (Cardoso et al., 2010).

Metagenomics may shed light in understanding the complex degradation routes of xenobiotics, which is currently poorly understood. These compounds can be toxic for living organisms and represent a threat to ecosystems. The use of molecular techniques based on genomic analysis can be applied in the monitoring of enzymes associated with the metabolism of xenobiotics, including herbicides and other pollutants (Malik et al., 2008).

Microbial communities can be screened by gene-specific PCR to detect the presence of genes of interest within a community. This method, however, has a bias of favoring previously known genes. An alternative method, which allows for the identification of novel genes and metabolic pathways, is the construction of expression libraries from metagenomic DNA. Through this method, positive clones expressing randomly cloned environmental genes are screened for their capacity to metabolize a specific substrate by plating in media containing a particular compound. In this way, the clones that metabolize this compound can be selected for DNA sequencing.

By gene-specific PCR it is possible to quickly identify genes associated with biodegradation in an environmental sample and sometimes affiliate them with a specific taxonomic group. This approach is limited by the fact that the metabolic pathways to which xenobiotics are submitted can encompass several steps, requiring more than a single enzyme to be catabolized. Therefore, cloning the full set of genes would be required to obtain the desired phenotype, which is not always possible since those enzymes can be encoded by different genes spread throughout the genome.

An alternative procedure for novel gene discovery is by Stable Isotope Probing (SIP). SIP is based on the incorporation of stable isotope-labeled substrates into molecular biomarkers. Once labeled substrates are provided to a microbial community, those microorganisms capable of metabolizing this substrate (e.g. a pollutant molecule) are likely to incorporate labeled atoms in their DNA, RNA and protein molecules. Nucleic acids with labeled atoms can be extracted to retrieve genomic material exclusively from a community capable of metabolizing a desired substrate (Malik et al., 2008). Furthermore, the sequencing of complete genomes of microorganisms can also reveal a set of metabolic genes with important applications for the biodegradation process, representing another important tool for bioremediation strategies (Eyers et al., 2004).

However, metagenomic studies usually result in the production of a great amount of sequence data so an advance in the capacity of bioinformatics tools is expected to deal with it. The processing of metagenomic data for subsequent genetic/environmental analysis requires a series of bioinformatics tools since the abundance of sequences obtained, especially with next generation sequencing equipments, can no longer be processed manually. Considering the high number of bioinformatics tools that have been and are still being developed, this chapter will focus on those most used for microbial community analysis in water pollution studies.

Sequences obtained from environmental metagenomic projects are usually compared to local or global nucleic acid sequence databases to identify the relationship between them and previously obtained data. Popular databases include GenBank (http://www.ncbi.nlm.nih.gov), the RDP (http://rdp.cme.msu.edu), CAMERA (http://camera.calit2.net), SILVA (http://www.arb-silva.de) and others, some of them dedicated to particular taxa. General databases are more fitted for a broad analysis whereas databases dedicated to a specific taxonomic group may contain more rare and reliable sequences. DNA sequence databases can also be divided between curated databases and non-curated ones. While some databases accept all sorts of nucleic acid and protein sequences (non-curated), others are more restrictive and perform a pre-selection of deposited sequences leaving a lower but more reliable amount of sequences available.

When using DNA sequencing for taxonomic identification ribosomal genes are frequently used but there is no gene or genomic region that is a golden standard for such procedure. In fact, some sequences present high similarity levels with more than one gene or a single species. Usually a DNA sequence from a single gene can confer a trustful identification to a family or genus level, depending on the gene and size of the DNA fragment. However, there is still much debate regarding which portion of genome is reliable for identification of organisms to a species level and whether such a perfect region does exist.

Several sequence alignment tools are available for the comparison of protein and nucleic acid sequences. A reliable alignment is required for a common practice within metagenomic approaches: the construction of phylogenetic trees. These trees, which may be constructed using software such as MEGA and ARB (http://www.megasoftware.net; http://www.arb-silva.de), are widely used to show sequence diversity within samples and to determine how these sequences are related to each other in evolutionary terms. It is important to highlight that sometimes the trees should not be interpreted as a precise evolutionary model of the studied sequences but as mere representations of which sequences are present in a selected sample and how they are related to reference sequences and to each other.

Computational tools can be also used in the quantification of differences between distinct datasets. For example, the software LIBSHUFF (http://whitman.myweb.uga.edu/libshuff.html) generates homologue and heterologous coverage curves to compare gene libraries in order to determine if two sequence libraries are statistically distinct from each other. UniFrac can be used for comparing microbial communities through principal component analysis, allowing several datasets to be compared at once, which helps in the detection of distribution patterns in communities of microorganisms and to correlate them with environmental variants (Lozupone & Knight, 2005).

Correspondence analysis is a similar approach, which helps to quantify how much of the microbial diversity is explained by environmental variables, with major applications for pollution studies as it determines the extent to which microbial communities are affected by pollutants (Vieira et al., 2008). Concerning microbial community diversity, DOTUR can be used as a tool to assign sequences as operational taxonomic units (OTUs) (Schloss & Handelsman, 2005). In addition, this software can be used to generate rarefaction and collector's curves and diversity indexes. Altogether these features help to quantify how much of the microbial diversity is being covered within a collected sample, to estimate how much of biological diversity is being retrieved through a metagenomic approach from a sample, and to determine how much effort is still required to reach a full coverage of sequence diversity.

Bioinformatics tools are evolving towards packages which aggregate different softwares to facilitate analysis. MOTHUR has emerged as a powerful tool for comparison of microbial communities (Schloss et al., 2009), allowing several steps of bioinformatics from metagenomic analysis to be performed in a single platform. MOTHUR implements tools like LIBSHUFF, DOTUR and UniFrac improving the use of bioinformatics tools to non-specialists. A simple flowchart summarizing the several steps of metagenomic analysis and bioinformatics is shown in Figure 1.

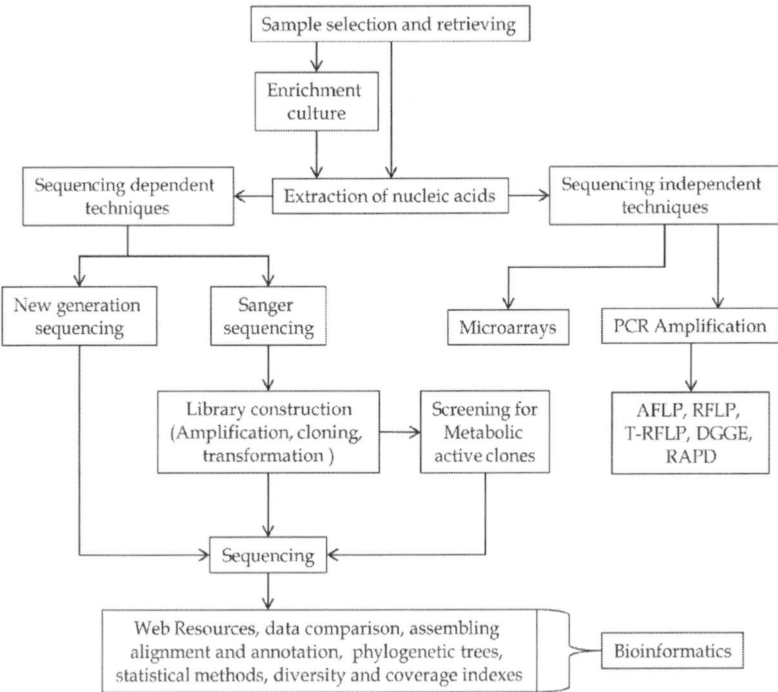

Fig. 1. Flowchart summarizing the several steps of metagenomic analysis and bioinformatics.

3. Metagenomics applied to water pollution studies

3.1 Guanabara Bay

Guanabara Bay is a coastal bay located in Rio de Janeiro state, SE Brazil, centered on latitude S 22°50' and longitude W 43°10', with a perimeter of 131 km and an area of 384 km², of which 328 km² are of water surface and 56 km² are comprised of islands. The mean water volume is 1.87 x 10⁹ m³. The bay measures approximately 28 km from west to east and approximately 30 km from south to north. The narrow entrance to the bay is 1.6 km wide. Guanabara Bay's drainage basin comprises a set of 32 separate sub-watersheds and is drained by 45 rivers and stream channels spread throughout an area of 4080 km². The bay's water surface area has suffered a reduction of about 10% due to urban expansion.

The bay's climate is tropical humid, with wet warm summers and dry cool winters, a mean annual air temperature of 23.7 °C and a mean rainfall of 1173 mm. The bay is characterized by high salinities and temperatures; the mean salinity is 29.5 ± 4.8 % with a total range from 9.9 to 36.8%, characterizing the bay as an estuarine environment. The average temperature is 24.2 ± 2.6 °C with a total range from 17.0 °C to 31.0 °C. Salinity decreases horizontally from the ocean entrance towards the inner reaches of the bay due to freshwater discharge while temperature increases from the ocean entrance towards the inner reaches (Paranhos & Mayr, 1993).

Guanabara Bay is located in a region that was originally covered by the Atlantic rainforest, an ecosystem of which only a very small fraction was conserved. Vegetation was cut down along the centuries as the region developed economically to become the second largest metropolitan area of Brazil. Most of the vegetation preserved along the basin is restricted to conserved portions of mangrove wetlands, in small fringes of the bay and covering mountain slopes. Studies have demonstrated a possible role for mangrove ecosystems in the attenuation of the anthropogenic impact imposed over the bay, raising awareness for the need to preserve and restructure those environments (Gomes et al., 2008; Santos et al., 2011). Guanabara Bay is surrounded by the metropolitan region of Rio de Janeiro city, encompassing several towns and smaller communities, totaling around 10 million inhabitants. Sewage runoff is discharged, mostly untreated, directly into the bay and in the rivers and streams that flow into it, which have been shown to undergo a severe eutrophication process (Mayr et al., 1989).

Rio de Janeiro went through an abrupt industrialization and intense population increase in the 1950s when the bay started to suffer with a severe anthropogenic impact. It is estimated that nowadays 540 tons of high biochemical oxygen demand waste and 5.5 tons of garbage are deposited daily into the bay (FEEMA, 2008). A large number of industries are present on its borders and thousands of ships circulate in its waters every year. Also, the bay's drainage basin possesses several urban, cattle raising, aquaculture, and agricultural areas in its surroundings, besides harboring one of the main industrial poles in the country. The bay is also subjected to several types of pollutants such as heavy metals, pesticides, antibiotics, oil, polycyclic aromatic hydrocarbons and organochlorines, raising the bay as one of the most impacted coastal environments in Brazil. Bay pollution has implications in public health issues, since its waters are utilized for recreational purposes and commercial fishing.

In the beginning of the 1990s a cleansing program was created for Guanabara Bay. Until 2010, almost a billion US dollars have been spent in an attempt to decrease the levels of pollutants discharged into the bay, which nonetheless persists as a highly impacted estuarine environment. Recent studies have shown that the bay has a high overload of organic and inorganic nutrients. Higher mean nutrient concentrations are found in the inner bay margins, due to sewage runoff and less efficient water renewal. At these same sites significant differences are observed in nutrient concentration between surface and bottom waters while nutrient levels are strongly reduced towards the bay's entrance (Vieira et al., 2007; Vieira et al., 2008).

Organic matter can either be deposited into the bottom sediments or undergo bacterial decomposition, consequently increasing nutrient availability in the water column and promoting eutrophication. The innermost reaches of the bay present elevated levels of biochemical oxygen demand as a result of the consumption of dissolved oxygen for degradation of organic matter derived from sewage runoff. High levels of chlorophyll a are also highest in the inner portions of the bay, presumably due to elevated nutrient overload allowing for higher phytoplankton production.

3.1.1 Metagenomics studies in Guanabara Bay

Estuaries can be biologically very diverse due to the mixing of seawater and freshwater so more knowledge about the composition and the ecological roles played by estuarine microbial communities is necessary. In 2007 and 2008, Vieira and colleagues studied the free-living planktonic Archaea and Bacteria diversity through the construction of 16S rRNA gene libraries, in a transect along Guanabara Bay, extending from a heavily polluted inner channel to the pristine costal seawater of the southern Atlantic Ocean. Nutrient levels along the transect revealed a clear trophic and pollution gradient along the bay, with higher nutrient levels in inner bay sites that decreased towards the bay`s entrance and beyond. This gradient reflected in bacterial abundance and production, which peaked in inner sites. Bacterial diversity in three of the sampled sites revealed clones associated with marine, estuarine and brackish waters and also with species present in sewage, indicating the effect of pollution in structuring the microbial communities of the bay. Additionally, in the highly polluted anoxic inner channel, several clones were affiliated with opportunistic pathogenic genera. This finding illustrates the risks to public health associated with impacted environments.

Exploring the diversity of bacterioplankton communities in a latitudinal gradient along Latin America through pyrosequencing of the V6 hipervariable region of the 16S rRNA gene, Thompson and colleagues, revealed in 2011 that among 7 different coastal seawater sites, that included Guanabara Bay, *Proteobacteria* followed by *Cyanobacteria*, *Bacteroidetes* and *Actinobacteria*, are the most abundant phyla. *Proteobacteria* accounted for more than 75% of the 17631 sequence tags obtained from the bay, with similar proportions observed in the other sites, indicating that this phylum is dominant among coastal seawater environments. Potentially pathogenic microorganisms were also detected within the bay, which is probably a result of nutrient overload that promotes the growth of heterotrophic bacteria, which include a diverse set of opportunistic pathogens (Thompson et al., 2011). This explains how a previously less harmful microbiota from a pristine aquatic environment is replaced by a set of new microorganisms that present higher risks to human health upon pollution impacts. This work is an excellent example of the applications of pyrosequencing to uncover rare members and perform a quantitative analysis of microbial communities which can be used to determine distribution patterns of microorganisms along the globe. On the other hand, the low extension of V6 tags obtained in this approach (~60 nt), hampers a full taxonomic characterization of sequences to a species level.

Archaeal species diversity in water samples from 4 distinct bay sites, with different levels of anthropogenic impact, was evaluated. Denaturing gradient gel electrophoresis (DGGE), revealed that free-living and particle attached archaeal communities are significantly different (Vieira et al., 2007). Suspended particulate material (SPM) is abundant in several sites of Guanabara Bay and SPM has been considered a hotspot of microbial activity because particles function as sites of intense heterotrophic metabolism (Azam et al., 2001). Interestingly, the highest levels of archaeal diversity found by Vieira and colleagues were detected in sites located in the interface between sewage polluted freshwater and coastal seawater, indicating that water mixing patterns promote increases in archaeaplankton diversity. Clones obtained through 16S rRNA gene library construction were mostly affiliated with other uncultured environmental Archaea. Libraries were dominated by typical estuarine *Euryarchaeota* followed by *Crenarchaeota* but in the anoxic heavily polluted inner channel library, most clones were associated with *Thermoplasmatales* and anaerobic methanogenic archaeal groups. The libraries also presented an abundance of species known for their hydrocarbon degradation activities, revealing a diverse set of microorganisms with potential applications for the bioremediation of ecosystems subjected to oil contamination.

Clementino and co-workers analyzed the archaeal diversity among several sites that included the bays water, agricultural soil, wastewater treatment plant water, halomarine sediment and a landfill leacheate, located in the bay's surroundings. 16S rRNA gene library clones revealed that wastewater samples were exclusively of the *Euryarchaeota* phylum, most of them affiliated with methanogenic Archaea. These high levels of pollutants in inner bay anoxic environments are promoting the growth of anaerobic Archaea with methane producing metabolism. Bay water samples presented clones of the *Euryarchaeota* and *Crenarchaeota* phyla affiliated with marine water samples (Clementino et al., 2007). It is likely that more archaeal phyla are present in this environment than could be detect by the low number of sequences obtained.

In 2010, Turque and colleagues published a work comparing the diversity of archaeal communities associated with the marine sponges *Hymeniacidon heliophila*, *Paraleucilla magna* and *Petromica citrina*, from Guanabara Bay with those from an offshore less impacted environment, the Cagarras Archipelago. The diversity of the *amoA* gene was also analyzed. This gene belongs to the operon *amo* that encodes the catalytic subunit of ammonia oxygenase, which catalyzes an ammonia oxidation reaction being therefore extremely relevant in the nitrogen biochemical cycling with potential applications in bioremediation processes. Results showed that although the two sampling sites are located very close to each other, Guanabara Bay waters presented higher levels of NH_3, total phosphorus, chlorophyll a, prokaryotic abundance and production when compared to the Cagarras Archipelago. No operational taxonomic units (OTUs) were shared between water samples of the two sites, indicating that differences in the levels of pollutants drastically change the archaeal communities of those environments. Some OTUs were also related to the archaeon *Methanoplanus petrolarius*, a species associated with petroleum contaminated anoxic environments, indicating a possible contamination of Guanabara Bay by petroleum spills. Another relevant finding was that sponges of the same species separated by a short distance presented distinct archaeal communities suggesting that environmental conditions can shape these sponge associated microbial communities. The *amoA* sequence diversity within sponge associated samples indicated that this gene can play a central role in the adaptation of sponges to eutrophicated environments, which usually present high levels of ammonia. This raises the hypothesis that sponges harboring a microbiota more capable of performing detoxification of their tissues would be more fitted to survive in an environment with high ammonia levels such as Guanabara Bay (Turque et al., 2010).

In 2011, Gonzalez and colleagues published the analysis of *amoA* and *nifH* gene diversity from bacterial communities Guanabara Bay water samples. The *nifH* gene encodes one of the nitrogenase protein complex subunits present in diazotrophic microorganisms, capable of reducing atmospheric nitrogen to ammonia. Most *amoA* sequences were related to uncultured environmental organisms, which probably correspond to new species that participate in ammonia oxidation. A similar pattern was observed for the *nifH* gene. The diversity of nitrogen fixing and nitrifying activity genes of these bacterial communities brings to light a complex set of microbial metabolic routes associated with nitrogen biochemical cycling in Guanabara, which can be extrapolated to other coastal estuarine ecosystems. Future studies are still necessary for a better characterization of the pollution impact of nutrients in this and other polluted ecosystems to expand the knowledge in the dynamics of urban estuarine biochemical cycles to understand how these impacted environments can function to remediate themselves (Gonzalez et al., 2011).

Mangroves are ecologically important ecosystems located in the interface between terrestrial, freshwater and seawater environments. Guanabara Bay harbored an extensive mangrove system which suffered significant reduction along the years due to anthropogenic impact. Very little is known about the microbial biodiversity in these habitats.

A 16S rRNA gene DGGE study of the bacterial communities in sediments from three distinct sites in Guanabara Bay's mangrove system, all subject to distinct levels of heavy hydrocarbon pollution, revealed that the bacterial microbial communities from the three sites were significantly different. This provides evidence that each mangrove site harbors distinct bacterial community patterns, which could be the result of the different levels of hydrocarbon contamination that each sampled site is subjected to (Gomes et al., 2008). When some of the DGGE bands were sequenced a diverse set of species that are capable of metabolizing hydrocarbons was revealed, illustrating the potential for bioremediation of environments subjected to oil spills.

Sediments of three mangrove forests from Guanabara Bay presenting distinct levels of PAH contamination (one of them near a petrochemical refinery) were analyzed regarding the diversity of genes encoding the multicomponent enzyme system naphthalene dioxygenase (NDO) that initiates the degradation metabolism of low molecular weight polycyclic aromatic hydrocarbons (PAH),). They revealed that PAH pollution is capable of shaping NDO gene diversity within these microbial communities (Gomes et al., 2007).

Hydrocarbon degrading bacterial communities from Guanabara Bay mangrove sediments were characterized utilizing mesocosm systems by Brito and colleagues in 2006. In this study, bacteria obtained from bay's mangrove sediments were inoculated into *in situ* 700 cm² surface mesocosms systems installed in the Guapimirim mangrove. Mesoscosms received 350 ml of petroleum and after 130 days of incubation bacteria retrieved from the mesocosms were isolated and inoculated in culture media containing one of five different hydrocarbons (octadecane, pristane, naphthalene, pyrene or fluoranthene) as a sole carbon source. Measurements of hydrocarbon degrading activities revealed that several isolates were capable of degrading hydrocarbons and some of them presented degrading activity of up to four kinds of aliphatic or aromatic hydrocarbons (Brito et al., 2006).

Sequencing of 16S rRNA genes from isolates showed that bacteria retrieved belonged to the following groups: *Gammaproteobacteria*, with a large number of isolates associated with genera known for their hydrocarbonoclastic activities; *Alphaproteobacteria*, of which most of the bacterial isolates were affiliated with hydrothermal vent strains, raising the possibility that these environments may harbor organisms with potential oil degrading activities; and *Actinobacteria*, which were affiliated with genera that due to their metabolic flexibility are capable of degrading different kinds of materials, like rubber, oils and hydrocarbons. More studies are still required to describe how microbial communities respond to hydrocarbon pollution in mangroves and other environments, especially regarding archaeal groups which present an unexplored set of metabolic activities.

These findings described above suggest that Guanabara Bay microbial communities are strongly affected by the high levels of anthropogenic impact to which this environment is subjected. The input of organic and inorganic nutrients promotes shifts in the diversity and abundance of microbial species that, in turn, are involved in biogeochemical cycles.

Different sites from Guanabara Bay present distinct levels of bacterial production and abundance (Vieira et al., 2008), which can be associated with the impact levels that each of those sites is subjected. To further understand how pollution affects the metabolism of these microbial communities a better understanding of the complex biochemical pathways that undergo in the Bay is still necessary therefore new generation sequencing techniques may play an important role in the process of expanding knowledge in this area in the near future. The degradation of Guanabara Bay and other aquatic ecosystems seems to promote the emergence of opportunistic pathogenic species, suggesting that pollution of aquatic environments represents a direct threat to public health and stresses the importance of preserving ecosystems for maintaining the quality of life of humans and other species in the planet. Opportunistic pathogens are common in polluted environments but a further description of pathogenic bacteria in the bay is still required, with special attention to antibiotic resistant bacteria, which endanger the millions of people living in the bay's surroundings. The microbial communities of the Bay's aquatic and surrounding mangrove ecosystems are repeatedly affected by hydrocarbon pollution due to nearby petrochemical facilities. Coincidently, the resident communities at these same sites seem to harbor the metabolic machinery necessary to remediate those impacts, increasing the relevance of studying their diversity more deeply. Hydrocarbon pollution is a global ecological issue and metagenomics can provide insights into the effects of hydrocarbon contamination on microbial communities worldwide and consequently help in the development of bioremediation strategies to bypass this common kind of environmental degradation.

3.1.2 Worldwide metagenomic studies of polluted habitats

Metagenomics has been applied to several studies of aquatic pollution worldwide to understand how the presence of pollutants affects microbial community composition and ecology. An example is the characterization of archaeal and bacterial sediment communities from two distinct portions of Western Europe's largest freshwater reservoir, Lake Geneva in Switzerland (Haller et al., 2011).

The Bay of Vidy is the most contaminated area of the lake: it is subjected to discharges from a wastewater treatment plant and shows the higher levels of nutrients and heavy metals while the Ouchy area is a nearby less polluted site. The sites were compared based on 16S rRNA gene library constructions from environmental DNA. Rarefaction analysis showed a higher number of bacterial OTUs in the Ouchy site, suggesting a higher diversity. Most of the retrieved bacterial sequences were associated with *Proteobacteria* and *Bacteroidetes* at both sites. However, phylogenetically distinct OTUs from these two bacterial groups were found at each site. A Multiple Factor Analysis (MFA) indicated that differences in bacterial community composition were statistically correlated to differences in the levels of organic matter, nutrients and heavy metals. All archaeal sequences belonged to the Euryarchaeota division. Several clones from the Vidy site were associated with archaea that present methanogenic metabolism, which can be associated with organic matter degradation in this polluted site. This work showed that microbial communities from nearby sites can suffer compositional changes due to treated sewage contamination, indicating not only that those communities are extremely sensible to alterations in environmental conditions but also that even after treatment sewage drastically affects species diversity within aquatic ecosystems.

Huang and colleagues (2011) published a metagenomic analysis of heavily polluted small streams in China. As those environments seem to be more sensible to pollution than large masses of water, the description of how they are affected by anthropogenic impacts is extremely relevant. The authors performed a DNA profiling technique based on Terminal Restriction Fragment Length Polimorfism (T-RFLP) in parallel to construction of bacterial 16S gene libraries to analyze three distinct streams subjected to industrial, agricultural and urban pollution. T-RFLP analysis showed significant differences in community composition between the three sites, even though physicochemical parameters were very similar among them. The results suggest that a large number of other factors that go beyond nutrient levels may be responsible for shaping the composition of the bacterial communities (Huang et al., 2011).

Gene libraries showed that *Betaproteobacteria* were widespread through all the three streams, although divisions such as *Alpha* and *Gamma-Proteobacteria*, *Bacteroidetes* and *Cyanobacteria* had distinct distribution patterns between the three samples, suggesting that the proportion between these taxa varies between streams. Furthermore, clones affiliated with several genera that have been previously associated with polluted environments such as *Flavobaterium*, *Rhodobacter* and *Hydrogenophaga* were retrieved. In addition to an evaluation of 16S rRNA gene libraries, this study presents a sequencing independent metagenomic analysis based on T-RFLP using environmental DNA. This technique can be applied for a crude characterization of microbial communities, with potential applications for biomonitoring strategies of aquatic environments. However, sequencing based techniques provide a deeper understating of bacterial community composition even though they are more expensive and time consuming.

Contamination of aquatic environments by sewage originated pollution poses a threat to human health. Fecal bacteria may infect humans that consume contaminated water for recreational, feeding or drinking purposes. Although quantification of *Enterobacteria* has been extensively used for water quality analysis, many species of this bacterial group can colonize different host species. Therefore, this analysis provides poor information about the origin of pollution (i.e. whether its source is human feces or fecal bacteria of other animals).

Wéry and colleagues (2010) studied the human specific fecal bacteria originated from wastewater treatment plant effluents focused on the following groups: *Bacterioidales, Clostridiales, Bifidobacteria*, and the *Bacillus-Streptococcus-Lactobacillus* (BSL) cluster. Construction of V6 region of the 16S rRNA gene libraries from genomic material of effluents from five French wastewater treatment plants was performed. A set of specific primers to target the selected bacterial groups were utilized (Wéry et al., 2010). Besides library construction, an analysis based on Capillary Electrophoresis Single Stranded Conformation Polymorphism (CE-SSCP) was performed to characterize the profiles of the four bacterial groups between samples, which suggested an smaller diversity of *Bifidobacterium* within samples when compared to the three other bacterial groups. This can possibly be accounted for by the small number of species from this group that is part of the human gut microbiota.

 Sequences obtained were affiliated with bacteria originated from feces, although some groups of *Bacteroides* commonly found in the human gut could not be retrieved from environmental samples, probably because some species are not fitted to survive in the wastewater environment and are more adapted to a host associated lifestyle.

Comparisons of bacterial species diversity between the wastewater treatment plants showed that the species profile of *Bacteroides, Clostridiales, Bifidobacterium* and BSL cluster is very similar, suggesting that specific bacteria from those groups are ubiquitous in wastewater treatment systems effluents. To identify the putative source of bacterial 16S rRNA gene sequences obtained through metagenomics the authors utilized a bioinformatics tool to associate sequences retrieved from samples to previously obtained fecal samples and observed that all *Bifidobacterium* and *Clostridiaceae* were associated with database sequences of fecal source while the other two groups were affiliated with sequences of fecal and non-fecal source. Overall, the results indicate *B. adolescentis, B. caccae, L. pectinoschiza and H. filiformis* as potential indicators of human fecal contamination in aquatic environments.

Although the exclusivity of those bacteria as human gut associated organisms is debatable, this work provides insights into the development of molecular techniques based on metagenomics and bioinformatics analysis, for fecal source tracking on aquatic environments, which can shed light into the quantification of impacts caused by wastewater effluents to which environments are subjected.

Bacterial resistance to antibiotics has recently become a public health issue. Aquatic environments are ecosystems where bacteria can easily exchange antibiotic resistance genes through horizontal gene transfer, promoting the spread of antibiotic resistant bacteria (Martinez, 2008). Analyses conducted in the Seine River in France and in the Zenne and Scheldt rivers in Belgium, subjected to urban and industrial sewage discharge, were performed focusing on antibiotic resistant bacteria, revealing an alarming truth for public health. Those environments presented a diverse set of heterotrophic bacteria that possess antibiotic resistance, including genera of important human pathogens (Garcia-Armisen et al., 2010). The 16S rRNA gene libraries showed that among bacteria that presented multiple resistance to several antibiotics obtained from sewage-contaminated rivers, most of them were affiliated with the *Bacterioidetes* or *Proteobacteria* phyla, reflected by the isolation of multi-resistant bacteria.

Based on the abundance of sequences associated with pathogenic bacteria, the authors raised the awareness to the health risks associated with polluted rivers, which could function as hotspots of dissemination of antibiotic resistance and as environments that promote shifts in bacterial behavior towards a pathogenic state. Additionally, the analysis of a 16S gene library originated from water samples of the heavily impacted Zenne river, revealed the severe impact to which microbial communities undergo upon sewage contamination, suffering drastic compositional changes that may result in alterations in ecosystem functioning.

Heavy metal contamination is a common form of aquatic pollution worldwide. Those compounds are toxic to living organisms and also present mutagenic activity. Microbial communities are capable of interacting with these compounds, consequently affecting their availability and impact in the environment. Rastogi and colleagues (2011) studied the Couer d'Alene river, located in the United States as a model environment of severe metal contamination. Metagenomic analysis integrating gene library construction and a microarray based technique, the PhyloChip were used to describe microbial communities from this site (Rastogi et al., 2011).

Couer d'Alene River has suffered an impact of 125 years of acidic ore mining, which resulted in extremely high levels of metal in its waters, that include As, Cr, Cu, Ni, Pb and Zn. Ribosomal small subunit gene libraries revealed only a modest bacterial diversity in this environment, as suggested by rarefaction curves, but the PhyloChip approach was capable of covering a much larger portion of the bacterial diversity which reached 40 phyla while sequences retrieved by library construction retrieved only 6 phyla. Most OTUs originated from river sediment samples were associated with the Phylum *Proteobacteria*, followed by *Firmicutes, Actinobacteria, Bacteroidetes, Acidobacteria* and *Chloroflexi*. In addition, several other phyla with a smaller number of OTUs could only be detected by the PhyloChip. Several OTUs obtained in this study were affiliated with sequences obtained from samples subjected to heavy metal pollution but also with soil and aquatic samples that were not subjected to such impacts.

Additionally, several OTUs where associated with microorganisms whose metabolism is involved in processes of heavy metal bioavailability and mobilization. The authors also explored the diversity of the *amoA* and *mcrA* genes, associated with microbial ammonia oxidation and methanogenic metabolism, respectively, through library construction. Libraries produced a small number of OTUs evidencing that few species harboring such genes are present in this environment. While the *amoA* gene library was dominated by species of the ammonia oxidizing genus *Nitrosospira*, the *mcrA* gene library was dominated by methanogenic species of the genus *Methanosarcina*, evidencing an important role of archaea in the process of methane production within the river. Although little information could be retrieved from the *amoA* and *mcrA* gene libraries, they provide the first insights into the characterization of the roles of ammonia oxidizers and methanogenic microorganisms in this environment.

Interestingly, the *amoA* and *mcrA* genes are associated with very distinct environments. The former commonly occurs in strictly aerobic microorganisms while the latter is usually associated with anaerobic metabolism. So, the apparent paradox of both genes being retrieved in the same environments is still to be elucidated. This work is a good example of how distinct metagenomic techniques can complement each other. Since microarrays rely on previously known sequences it cannot be applied for detecting new species. Furthermore, this technique provides very little information concerning the composition of DNA sequences, revealing only if they are present or absent. On the other hand, the detection of novel taxa can be reached by 16S rRNA gene library construction. One limitation is that rare groups tend to be neglected by this approach, even with a very large sequencing effort.

Hydrocarbon pollution is a global issue, oil spills cause contamination of aquatic environments affecting living organisms in most areas of the world. Efforts to avoid, reduce, and remediate those continuous and/or acute contaminations are extremely necessary. Some advancement in this subject, however, has been recently achieved. For example, Marcos and colleagues (2009) identified bacterial populations in sub-Antarctic marine sediments, subjected to severe hydrocarbon pollution, capable of degrading polycyclic aromatic hydrocarbons (PAH). These findings revealed potential bioremediation strategies based on bacterial oxygenase genes obtained from those environments. In this work, samples were retrieved from the Ushuaia Bay, Argentina, a region subjected to constant small accidental hydrocarbon spillages, due to loading and unloading of petroleum-derived substances in a nearby pier. Using metagenomic DNA, construction of gene libraries were performed using primers designed for the gram-negative bacterial dioxygenase gene (Marcos et al., 2009).

The dioxygenase is the catalytic subunit of a multicomponent enzyme complex that catalyzes the insertion of molecular oxygen into benzene rings, an important step in PAH degradation pathways. Phylogenetic analysis of the dioxygenase protein sequences deduced from these gene sequences revealed a diverse set of enzymes in Ushuaia Bay's bacterial communities. Results revealed important biological indicators of PAH pollution and highlight the potential applications of metagenomics in the search for enzymes to be used for the development of PAH bioremediation strategies.

An recent evaluation of the bacterial diversity within mangrove sediments revealed several groups sensible to oil contaminations, which represent potential oil spillage indicator organisms in ecosystems (Santos et al., 2011).

In this work, bacterial communities in microcosms containing mangrove sediments exposed to 2% and 5% v/w of fuel oil were analyzed prior to oil exposure and after 23 and 66 days of oil contamination. Pyrosequencing of partial 16S rRNA gene sequences revealed that bacterial communities from all samples were dominated by the *Proteobacteria* phylum, mostly of the *Gamma*, *Delta* and *Alpha-proteobacteria* classes. Interestingly, petroleum contaminated microcosms presented higher indexes of species richness when compared to their control samples. Additionally, the authors observed significant shifts in bacterial genera abundance associated with hydrocarbon degradation (*Alcanivorax*) and hydrocarbon contaminated environments (*Marinobacterium, Marinobacter, Clostridium,* and *Fusibacter*). On the other hand, an intense decrease in the abundance of sequences associated with the genera *Helia* after hydrocarbon contamination indicated that this group seems to be very sensitive to oil pollution.

A similar study focused on the diversity of microeukaryotes retrieved from the same mangrove system within microcosms subjected to the same levels of oil contamination. DGGE and 18S rRNA gene library construction revealed organisms within the microeukaryote group that could be used as potential bioindicators in the monitoring of oil pollution (Santos et al., 2010).

Results suggested that Fungi and Metazoa, the originally dominant group in the mesocosm, are the most sensitive to oil contamination, being replaced by *Stramenopiles* as the most abundant group after oil contamination. Therefore, the balance between those two groups can reflect the level of oil impact to which an ecosystem is submitted. Interestingly microeukaryote species richness and diversity indexes were lower in the oil contaminated mesocosm, in opposition to the prokaryotic pattern of higher diversity and richness in contaminated mesocosm.

Determining the reason why these communities behave differently upon oil contamination will shed light on the variables that regulate microbial community composition on polluted environments and possibly establish new ecological relationships by which prokaryotic and microeukaryotic populations affect each other in pristine and polluted ecosystems. Also, the comparison of microbial communities in microcosms subjected to oil contamination in different time periods suggests that shifts in prokaryotic and microeukaryotic communities are time-dependent, undergoing distinct alterations according to the time passed since contamination. Altogether these results provide precious information to be used in the development of bioindicators based on microbial diversity, abundance and community composition. Even though those results are very promising, further study is required to determine if microbial communities respond to oil contamination in a similar pattern as they respond in microcosm environments before reliable biomonitoring strategies can be developed.

4. Conclusions

We now know that pollution severely affects microbial communities indicating how fragile aquatic biota can be, so it is necessary to explore how exactly those impacts affect microorganisms directly (e.g. characterizing which of pollutants cause death of organisms and which ones promote slowing of growth). Rarefaction curves generated from metagenomic analysis indicate that a large amount of the diversity of Bacteria and Archaea is not being covered, calling for further efforts to fully identify these organisms, especially when focusing on rare groups.

New generation sequencing techniques seem to be the option for a complete access to Earth's Biota. Also, due to the capability of those methodologies to generate very large datasets, they are more fitted for quantitative inference, which is only poorly explored through library construction due to the smaller number of sequences that is generated through this method when compared to these new generation sequencing technologies.

Future studies focusing on microbial communities from aquatic environments subjected to anthropogenic pollution, to determine their composition, structure, metabolic capacities, and ecological relationships are still necessary for a deep understanding of how those complex environments function and how the severe impacts to which they subjected can be reversed or at least attenuated. Function driven metagenomics, focused on specific genes associated with biogeochemical cycles (e.g. *amo* genes) contribute to a more detailed interpretation of the data gathered so far, so scientists can know not only which microorganisms are living in impacted environments but also how they are behaving biochemically. Furthermore, the utilization of metagenomic techniques will help to determine what effects pollution produces in aquatic microorganisms in a global level, specifying which alterations are common to all impacted environments and which ones are associated with a specific ecosystem, pollutant or a microbial group.

A great number of polluted sites where microbial diversity was explored through metagenomics are subjected to more than one kind of pollutant (industrial, wastewater, agricultural, etc.) Analyzing similar sites subjected to different pollutants will contribute to the understanding of which pollutants are more aggressive against environmental microbial communities and what sorts of alterations each one of them is capable of promoting.

For example, several studies suggest that while some sorts of pollutants, such as heavy metal and hydrocarbons, tend to alter or decrease microbial diversity, sewage contamination usually produces increases in microbial diversity. Sequences originated from polluted environments are often associated with pristine sources, indicating that those sites may harbor an important microbiota capable of surviving and growing in heavily polluted sites. Further exploring the diversity of organisms in these ecosystems will provide relevant insights for biomonitoring strategies as these microorganisms may present important metabolic traits that can be applied for bioremediation. Therefore, pristine environments are as much important as polluted ones in the development of biotechnology strategies, and thus require further efforts to elucidate their microbiota.

Certainly, an enormous amount of species, genes, proteins, enzymes, and metabolic pathways are still to be discovered. Guanabara Bay has been shown to be a potential site harboring those unexplored biological units, which presents potential applications for biomonitoring, bioremediation and even sanitary policies.

Metagenomic analysis showed that *Proteobacteria* are widely spread throughout polluted environments, therefore, further studies focused on this phylum will contribute to our understanding of the anthropogenic impacts in aquatic environments. The study of the behavior of microbial communities in Guanabara Bay during and after the cleansing program to which it is being submitted can provide insights into how aquatic life in heavily polluted aquatic environments responds to attempts to revert impacts, to determine if such an attempt is successful and to quantify the extent of the damage that can be reversed. Due to its characteristics, Guanabara Bay could be used as a model ecosystem for this sort of analysis. Therefore, Guanabara Bay has an unexplored biological and biotechnological richness requiring further research, specially focusing on microbial groups that are poorly studied such as environmental microeukaryotes, viruses and archaeas. For these future studies metagenomic approaches will surely be indispensable and will certainly contribute much valuable information.

5. Acknowledgement

This work was partially funded by Fundação Carlos Chagas Filho de Amparo à Pesquisa do Estado do Rio de Janeiro (FAPERJ) and Conselho Nacional de Desenvolvimento Científico e Tecnológico (CNPq).

6. References

Azam F & Long RA. (2001). Sea snow microcosms. *Nature*. 414:495-498.

Brito EM, Guyoneaud R, Goñi-Urriza M, Ranchou-Peyruse A, Verbaere A, Crapez MA, Wasserman JC & Duran R. (2006). Characterization of hydrocarbonoclastic bacterial communities from mangrove sediments in Guanabara Bay, Brazil. *Researh in Microbiology*. 157:752-762.

Cardoso AM, Vieira RP, Paranhos R, Clementino MM, Albano RM & Martins OB. (2011). Hunting for extremophiles in rio de janeiro. *Front Microbiol*. 2:100.

Cardoso AM, Clementino MM, Vieira RP, Cavalcanti JJV, Albano RM & Martins OB. (2010). "Archaeal metagenomics: bioprospecting novel genes and exploring new concepts" in Metagenomics: Theory, Methods, and Applications, ed. D. Marco (Wymondham: Caister Academic Press), 159-169.

Clementino MM, Fernandes CC, Vieira RP, Cardoso AM, Polycarpo CR & Martins OB. (2007). Archaeal diversity in naturally occurring and impacted environments from a tropical region. *Journal of Applied Microbiology*. 103:141-151.

Eyers L, George I, Schuler L, Stenuit B, Agathos SN & El Fantroussi S. (2004). Environmental genomics: exploring the unmined richness of microbes to degrade xenobiotics. *Applied Microbiology and Biotechnology*. 66:123-130.

FEEMA. (1998). Qualidade de água da Baía de Guanabara 1990/1997, *Fundação Estadual do Meio Ambiente*. pp. 100.

Garcia-Armisen T, Vercammen K, Passerat J, Triest D, Servais P & Cornelis P. (2011). Antimicrobial resistance of heterotrophic bacteria in sewage-contaminated rivers. *Water Research*. 45:788-796.

Gilbert GA & Dupont CL. (2011). Microbial metagenomics: beyond the genome. *Annual Review of Marine Science*. 3:347-371.

Gomes NC, Borges LR, Paranhos R, Pinto FN, Krögerrecklenfort E, Mendonça-Hagler LC & Smalla K. (2007). Diversity of ndo Genes in Mangrove Sediments Exposed to Different Sources of Polycyclic Aromatic Hydrocarbon Pollution. *Applied and Environmental Microbiology*. 73:7392-7399.

Gomes NC, Borges LR, Paranhos R, Pinto FN, Mendonça-Hagler LC & Smalla K. (2008). Exploring the diversity of bacterial communities in sediments of Urban mangrove forests. *FEMS Microbiology Ecology*. 66:96-109.

Gonzalez AM, Vieira RP, Cardoso AM, Clementino MM, Albano RM, Mendonça-Hagler L, Martins OB & Paranhos R. (2011). Diversity of bacterial communities related to the nitrogen cycle in a coastal tropical bay. *Molecular Biology Reports*. DOI: 10.1007/s11033-011-1111-9.

Haller L, Tonolla M, Zopfi J, Peduzzi R, Wildi W & Poté J. (2011). Composition of bacterial and archaeal communities in freshwater sediments with different contamination levels (Lake Geneva, Switzerland). *Water Research*. 45:1213-1228.

Huang Y, Zou L, Zhang S & Xie S. (2011). Comparison of Bacterioplankton Communities in Three Heavily Polluted Streams in China. *Biomedical and Environmental Sciences*. 24:140-145.

Lozupone C & Knight R. (2005). UniFrac: a new phylogenetic method for comparing microbial communities. *Applied and Environmental Microbiology*. 71:8228-8235.

Malik S, Beer M, Megharaj M & Naidu R (2008). The use of molecular techniques to characterize the microbial communities in contaminated soil and water. *Environment International*. 34:265-276.

Marcos MS, Lozada M, Dionisi HM. (2009). Aromatic hydrocarbon degradation genes from chronically polluted Subantarctic marine sediments. *Letters in Applied Microbiology,* 49:602-608.

Martinez, JL. (2008). Antibiotics and antibiotic resistance genes in natural environments. *Science.* 321:365-367.

Mayr LM, Tenembaum DR, Villac, MC, Paranhos R, Nogueira CR, Bonecker SLC & Bonecker (1989). Hydrobiological characterization of Guanabara Bay. In Coastlines of Brazil, eds. O. Magoon & C. Neves. American Society of Civil Engineers. Charleston July, 1989.

Rastogi G, Barua S, Sani RK & Peyton BM. (2011). Investigation of Microbial Populations in the Extremely Metal-Contaminated Coeur d'Alene River Sediments. *Microbial Ecology.* 62:1-13.

Paranhos R & Mayr LM. (1993) Seasonal patterns of temperature and salinity in Guanabara Bay, Brazil. *Fresenius Environmental Bulletin.* 2:647-652.

Santos HF, Cury JC, Carmo FL & Rosado AS, Peixoto RS (2010). 18S rDNA sequences from microeukaryotes reveal oil indicators in mangrove sediment. *PLoS ONE.* 5:12437.

Santos HF, Cury JC, Carmo FL, dos Santos AL, Tiedje J, van Elsas JD, Rosado AS & Peixoto RS (2011). Mangrove bacterial diversity and the impact of oil contamination revealed by pyrosequencing: bacterial proxies for oil pollution. *PLoS ONE.* 6:16943.

Schloss P & Handelsman J. (2005). Introducing DOTUR, a computer program for defining operational taxonomic units and estimating species richness. *Applied and Environmental Microbiology.* Vol.71, No.3 (May 2004), pp. 1501-1506, ISSN 0099-2240

Schloss P, Westcott S, Ryabin T, Hall J, Hartmann M, Hollister E, Lesniewski R, Oakley B, Parks D, Robinson C, Sahl J, Stres B, Thallinger G, Van Horn D, Weber C. (2009). Introducing mothur: open-source, platform-independent, community-supported software for describing and comparing microbial communities. *Applied and Environmental Microbiology.* 75:7537-7541.

Singh J, Behal A, Singla N, Joshi A, Birbian N, Singh S, Bali V & Batra N. (2009). Metagenomics: Concept, methodology, ecological inference and recent advances. *Biotechnology Journal.* 4:480-494.

Thompson FL, Bruce T, Gonzalez A, Cardoso A, Clementino M, Costagliola M, Hozbor C, Otero E, Piccini C, Peressutti S, Schmieder R, Edwards R, Smith M, Takiyama LR, Vieira R, Paranhos R & Artigas LF. (2011). Coastal bacterioplankton community diversity along a latitudinal gradient in Latin America by means of V6 tag pyrosequencing. *Archives of Microbiology.* 193:105-114.

Turque AS, Batista D, Silveira CB, Cardoso AM, Vieira RP, Moraes FC, Clementino MM, Albano RM, Paranhos R, Martins OB & Muricy G. Environmental Shaping of Sponge Associated Archaeal Communities. *PLoS ONE.* 5:15774.

Vieira RP, Clementino MM, Cardoso AM, Oliveira DN, Albano RM, Gonzalez AM, Paranhos R & Martins OB. (2007). Archaeal Communities in a Tropical Estuarine Ecosystem: Guanabara Bay, Brazil. *Microbial Ecology.* 54:460-468.

Vieira RP, Gonzalez AM, Cardoso AM, Oliveira DN, Albano RM, Clementino MM, Martins OB & Paranhos R. (2008). Relationships between bacterial diversity and environmental variables in a tropical marine environment, Rio de Janeiro. *Environmental Microbiology.* 10:189-199.

Wéry N, Monteil C, Pourcher A-M & Godon J-J. (2009). Human-specific fecal bacteria in wastewater treatment plant effluents. *Water research.* 44:1873-1883.

6

Mathematical Modeling of the Suspended Sediment Dynamics in the Riverbeds and Valleys of Lithuanian Rivers and Their Deltas

Alfonsas Rimkus and Saulius Vaikasas
Water Research Institute of Aleksandras Stulginskis University
Vilainiai, Kedainiai
Lithuania

1. Introduction

Flooded river valley meadows particularly in their deltas are very important for water quality and river ecology, as growing there grasses entrap the flow sediments and sediment bond chemical materials brought by river. Consequently the water of seas, gulfs and lagoons becomes much clearer. The river flow bring from the river-basin area the washed from agriculture fields ground, so called "wash load", which contain many fine clay particles with adsorbed organic and nutrient materials as nitrogen phosphorus or even heavy metals. The suspended sediment load in significant part is brought into the valleys, where its deposition is going on. So the water quality in the rivers and seas is improved.

It was estimated that in the Nemunas delta during the period 1950 – 1981 it got deposited about 250 t potassium, 950 t phosphorus, 38000 t of calcium, and 147000 t organic matter rich with nitrogen (Vaikasas et al. 1997). It did not get in the Curonian Lagoon and the Baltic Sea (Fig. 1). Nevertheless, the grass–covered floodplains and deltas are often separated from rivers by dikes for intensive agriculture or other purposes. When such systems are designed, it is necessary to model sediment deposition in these separated areas to estimate the future increase of water contamination of the rivers and their receivers: gulfs and seas. However, the calculation methods used in the known mathematical models are adapted for flow over the sandy river bottom only (Bixio & Defina, 2004). Calibration results of the mathematical model of the river Nemunas delta with common sediment deposition formulae for riverbed flows did not correspond to the data of measurements (Rimkus & Vaikasas, 1997). Sediment deposition in the grassed floodplains was several times greater than on the sandy beds, as in the meadows there are other boundary conditions much more favourable for sediment deposition. Therefore the calculated increase of water contamination would be much less than the possible real. Consequently, it was necessary to study the peculiarities of flow and sediment motion under these conditions as well as to work out the sediment deposition calculation formulae, suitable for calculations on the new created mathematical model.

Intensive sediment deposition in grassed floodplains was estimated and significant reduce of nutrients was found in many other rivers (Large & Petts, 1994; Fustec et al., 1991; Haycock & Burt, 1990, 1991; Middelkoep & Haselen, 1999). The decrease of sediment-bound phosphorus and nitrogen reached even 80-87% (Jankowska, 2006; Lamsodis & Vaikasas,

2005). It was also stated that accurate prediction of the movement and deposition calculation of muddy sediments is highly desirable, however knowledge of these complex objects is limited, and there is no generally accepted formula for accurate calculation of sediment transport rate (Soulsby et al., 2010).

The focus of our studies was processes of suspended sediment deposition in the flooded delta of the river Nemunas and in the grassed floodplain of the river Nevėžis and the small river Virvytė.

Investigations of sediment motion analysis in the Nemunas delta showed, that the suspended sediment deposition in flooded valley meadows was very intensive. It reaches about 80% of summary sediments bought to the delta by river flow. That is due to the favourable conditions for sediment deposition in the delta. The river flow of the Nemunas in its delta is divided in two parts: the part overflowing in the wide valley and the part remaining to flow further along the riverbed. During the floods, particularly the large ones, flow velocities in the riverbed below this overflow in the valley are hardly decreased; therefore the sandy sediments are intensively deposited here. They pond the water level higher and increase the overflow into the valley, where the deposition of fine sediments is hardly increased also.

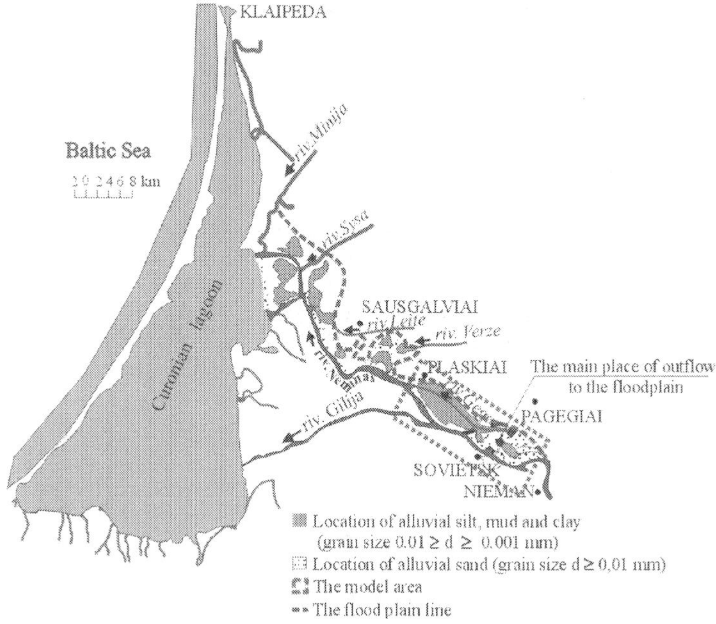

Fig. 1. Scheme of floodplain of the Nemunas delta and the Curonian Lagoon (The model area is indicated by a rectangle)

After the maximal water discharges of floods and between them, the deposited large amounts of sediments are washed off and leveled. This process has been modeled in the Nemunas delta for 50 years with various flood intensities. Consequently the dynamics of sediment motion in the riverbed and valley flows has been investigated.

In the flooded meadows of the river Nevėžis the suspended sediment deposition in grassed areas has been investigated (Vaikasas, 2010). The received data were employed for calibration of new formulae for the calculation of sediment deposition in grassed valleys of plane Lithuanian rivers.

In the small river Virvyte the sedimentation process in the riverbed and in flooded valley has been investigated under the conditions, when the cascade of hydropower plants along the whole river was build, which led to the modified conditions for water flora and fauna life. The hydro energetic plants worsened the ecological conditions of river, therefore the coordination ways of environmental and water energy employing needs were studied.

2. Methods of investigation

The sediment deposition in the riverbeds and floodplains in investigated objects has been estimated by calculation of deposited sediment amounts for the period 1950 - 1991, for which the hydrometric data about the sediment concentrations was available. During this time of investigations the periods with high and low floods were observed, then on the bottom of riverbed the sediments were either accumulated from year to year or washed. Subsequently the river became either more shallow or deeper, also the water levels during the floods were changing accordingly. The discharge of water overflowing into the valley depends on these water levels. Consequently during these periods the sediment deposition in the valley was also changing. By calculations for long period the average data of sediment deposition have been received.

According to calculations, during the high floods the riverbed below the places of water overflow in the valley was filed with sediments almost fully, as the flow velocity decreased there significantly, and the flow was not able to bring the sediments further. Such were the floods in 1951, 1958 and 1979. The water overflow into the valley and sediment deposition was very increased then. During the sinking of flood and after it the accumulated sediment layer in this strip was quickly washed out and spread below. The water flow remains normal.

Sediment deposition calculations were performed for 4 sediment fractions found by investigations. Their particle diameters were 0.005, 0.01, 0.02 and 0.1 mm. These are the sediment particles from clay to fine sand. According to investigations of hydrometric stations the concentrations of these sediment fractions fluctuated in large diapason – from 3 to 100 mg/l. Their measured meanings were employed for calculations.

Sediment deposition intensity depends on the flood size. During low floods when small water discharge flows; all sediments brought into the valley get settled. With increasing floods the certain part of sediments is brought to the end of investigated valley interval and returned to the riverbed. Therefore for estimation of this process the investigations were necessary to be continued in long term period.

Sediment deposition in the Nemunas valley was calculated applying our own hydraulic-mathematical model "DELTA" created for the study of the Nemunas delta (Fig. 2) (Rimkus et al., 2004, 2007). The known mathematical models (MIKE 21 1995) were not applicable for our purpose because they are not adapted for sediment deposition calculations in the flooded meadows, for which the special formulae are to be applied.

For riverbed flows it is characteristic the bottom sediment load with high sediment concentration. The sediment deposition process begins when the flow in this bottom sediment layer is saturated, i.e. when the transportable sediment concentration or critical flow velocity is achieved. The concept of critical velocity or transportable concentration cannot be applied for sediment deposition calculations for flow with grass-covered bottoms, where neither the bed load of sediments nor high concentrations necessary for flow saturation are observed. The unique sediment deposition is constantly going on at the bottom in the flow over grass. Having measured the water turbidity, it was determined that at the grass level suspended sediment concentration exceeded the average concentration by only about 1.2 times, while the concentration near the sandy bottom of the river Minija (tributary of the Nemunas River in its delta) was eight times higher (Vaikasas & Rimkus, 1996). Consequently, the boundary conditions for sediment deposition in riverbed on grassed flood plains are quite different. Therefore the calculation equations, in both cases must be different as well.

The ability of grasses to entrap the sediment was already noticed earlier (Barfield et al., 1979, Thornton et al., 1997; Pasche & Rouve, 1984; Cristiansen & Wiberg, 1997; Carpena et al., 1999; Deletic, 2001). However the sedimentation process in grassed food plains was yet not investigated properly.

Method for calculation of sediment deposition in the grass-covered floodplain, proposed by Rimkus, was created with estimation of grass ability to entrap the sediments (Rimkus et al., 2007). Because of the low flow velocity between the grasses, the sediment deposition in them becomes similar to the deposition in still water. It is proportional to the fall velocity of sediment particles and on the sediment concentration between the grasses, which is formed by concentration in the flow at the grass layer. Therefore the sediment deposition into the unit of bottom area can be expressed as follows:

$$D = k_{cor} w C_b \tag{1}$$

where w – the fall velocity of sediment particles, C_b – sediment concentration at the flow bottom, i.e. at the surface of grass layer, k_{cor} – correction coefficient depending on the state of grasses; for the luxuriant grass it is greater.

The fall velocity of sediment particles depends mostly on their diameter. To estimate it, the composition of sediment particles must be known; therefore the water examples containing suspended sediments are taken during the floods. In natural water samples, fine sediments usually make the aggregates. During the laboratory experiments, the aggregates commonly are destroyed, and the physical sediment composition is received. However for cohesive sediment deposition calculations the deposition of aggregates must be estimated, therefore the experiments of sediment composition for this aim is to be performed of natural water samples with non-destroyed aggregates. Then the aggregates receive the equivalent diameter of sandy particles. It gives the real fall velocity of aggregates (Pukštas & Vaikasas, 2005).

Usually sediment concentration in the flow is expressed by average concentration $\overline{C}$; therefore it is necessary to estimate their ratio $F = \overline{C} / C_b$. Then formula (1) changes so:

$$D = k_{cor} w \overline{C} / F \tag{2}$$

For calculation of ratio F the next formula was derived:

$$F = \left(\frac{a}{h-a}\right)^z \left[\int_a^h \left(\frac{h-y}{y}\right)^z v_y dy\right] \cdot \frac{1}{\int_a^h v_y dy}, \quad z = \frac{w}{\beta k u_*}. \tag{3}$$

where h – water depth, y - distance of investigated point from the bottom, v_y – water velocity at the distance y from the bottom, $a=0.3h_{gr}$, h_{gr} – thickness of grass layer, $k=0.4$ – Van Karman number, z – Rouse number, β - ratio of the sediment and momentum diffusion coefficients, u_* - shear velocity.

For calculation of ratio F, the velocity distribution along the water depth is necessary. This distribution can be accepted, for example, as logarithmic. The vertical velocity distribution depends on the flow turbulence distribution, which depends on the boundary conditions. In the width valleys this conditions are simple and similar to the ones existing in wide enough experimental channels with grass-covered bottom, in which the logarithmic distribution of velocities is found (Kouwen & Unny, 1973; Christensen, 1985; Temple, 1986; Kouwen, 1987; Yurchuk, 1999). Such velocity distribution was used in our model. For fine clay sediments, it is received F=1.

This sediment deposition calculation method is described in detail in the published articles (Rimkus et al., 2007; Rimkus & Vaikasas, 1999). Sediment deposition calculations were performed according these formulae.

To calculate sediment motion in the Nemunas riverbed it was necessary to chose the formulae for calculation of discharges of bottom and suspended sediments. Difficulty was, that mostly they were derived and therefore are suitable for sediment particles coarser than 1 mm, while in the Nemunas Delta they are finer. The formulas of van Rijn (1993) was chosen, which are suitable and for particles with diameter equal to 0.2 mm. For bottom sediments they are as follows:

$$q_{dg} = 0.153(s-1)^{0.5} g^{0.5} d_{50}^{1.5} d_*^{-0.3} T^{2.1} \text{, when } T = (\tau - \tau_{kr})/\tau_{kr} < 3, \tag{4}$$

$$q_{dg} = 0.1(s-1)^{0.5} g^{0.5} d_{50}^{1.5} d_*^{-0.3} T^{1.5} \text{, when } T \geq 3, \tag{5}$$

$$\tau = \rho g \left(\frac{v_{vid}}{18 \lg(4h/d_{90})} \right)^2 \tag{6}$$

$$\tau_{kr} = (\rho_s - \rho) g d_{50} \theta_{kr} \tag{7}$$

$$d_* = d_{50} \left((s-1) g / v^2 \right)^{1/3} \tag{8}$$

$$s = \rho_s / \rho \tag{9}$$

here q_{dg} – the bed load; d_{50} and ir d_{90} – diameters of sediment particles of probability 50 and 90 %; τ – the bed shear stresses; τ_{kr} – critical value for sediment deposition of bed sear stresses; v_{vid} – the average velocity of flow; h – water depth; d_* – non dimensional diameter of sediment particles; ρ_s and ir ρ – density of sediments and water; v – viscosity of fluid.

For the suspended sediments van Rijn (1993) proposed such formulas:

$$q_{sknd} = F v_{vid} h C_a \tag{10}$$

$$C_a = 0.15 \frac{d_{50}}{a} \frac{T^{1.5}}{d_*^{0.3}} \tag{11}$$

$$a = \Delta / 2, \tag{12}$$

here q_{sknd} – the discharge of suspended sediments; F – ratio of the sediment concentration, existing at the height a, and the average in the flow concentration, which can be estimated according the formula 2; Δ – the height of sandy waves at the bottom.

The calculation of suspended sediment deposition within the regions of investigated flow requires estimation of the stream velocities, which is most often done by using one-dimensional calculation methods. However, these methods determine only an average flow velocity and the sediment discharge in the floodplain as a whole. According to our model, created for estimating the sediment distribution across the valley, the river flow is divided into several strips with equal water discharges, and one-dimensional equations are employed in calculations. (Fig. 2). Consequently, the model turns to a quasi two-dimensional one and, therefore, it can give more exact results. The application of real 2D models would be very difficult, since a large number (several millions) of net points must be chosen due to a complicated valley relief, and the calculations take too much time. The work with such models is not efficient particularly, when the flow and sediments discharges are variable. Therefore, our quasi-2D model was used. This enabled us to calculate the sediment deposition in a many-year period and to estimate thoroughly the influence of HPP ponds and weir heights on the quality of river water as well.

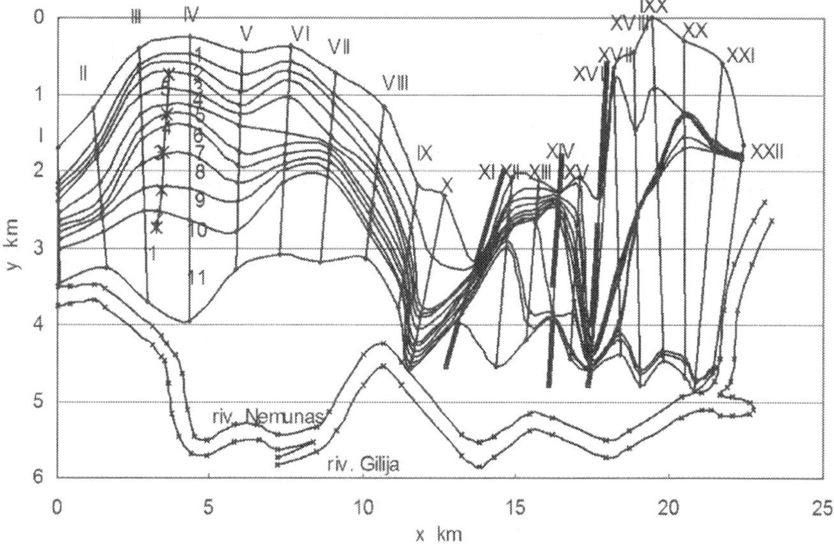

Fig. 2. Distribution of computerised stream strips at the investigated section of the Nemunas delta model (the flood of 1996): I-XXII – cross-section numbers; 1-5 – numbers of measurement posts; 1-11 – numbers of flow strips.

3. Modelling of the suspended sediment dynamics

3.1 Sediment motion dynamics in the river Nemunas delta

Amounts of sediments for four fractions deposited in upper and down strips of Nemunas delta calculated according the described calculation method are plotted in Fig. 3 and 4.

The change of sediment layer in the riverbed during 1950-1991 is represented in Fig. 5. There are plotted the thickness of sediment layer accumulated till this time. They were calculated for 4 segments. The first two are in down strip of valley and the third and fourth – in the upper one.

As the calculation results show, most of the sediments in the valley and also in the riverbed was deposited at the beginning of investigations, as the floods were large then (Rimkus &Vaikasas, 1999, 2010). Particularly intensive deposition was in 1958, when the flood was very large (probability 1%).

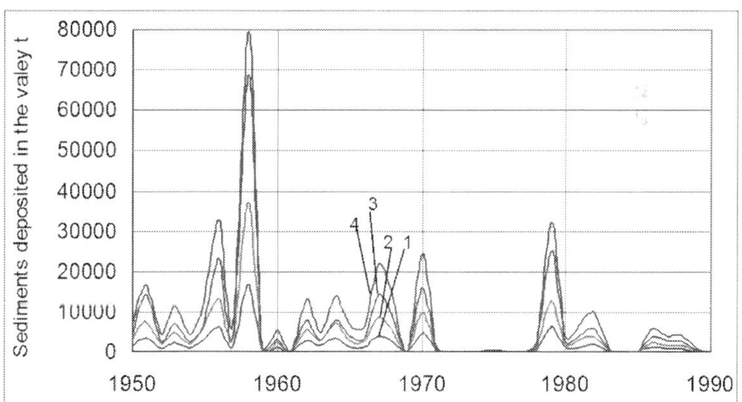

Fig. 3. Sediments deposited in down strip of valley: 1 – diameter of fraction particles 0.005 mm; 2 – diameter of fraction particles 0.01 mm; 3 – diameter of fraction particles 0.02 mm; 4 – diameter of fraction particles 0.1 mm. The summary amount of sediments settled during 1950–1991 time interval is equal to 739000 t.

The conditions for sediment deposition in the down and the upper parts of the delta are different, so the settled amounts of sediments are different too. The valley in the upper part is much wider; therefore the sediment deposition was more intensive there. However the floods overflow more rare in this part, because the altitudes of the overflow the places are rather high, therefore it decreased the overflow rates and sedimentation amount. In the upper strip the river did not overflow in 19 years of 42 years of the entire investigations. In the down strip area the river did not overflow only 11 times, as the overflow places are lower here. That increased the sedimentation here. However because of the less area of this strip, the sediment deposition was less intensive there.

The deposited amounts of sediments are quite large. In the upper part of delta it was deposited 35% of all sediments brought by river. Sedimentation in both strips reaches 60% of the brought by the river flow. Summary amount of sediments makes 2.3 mln t. The sediment deposition is going on also and in valley part near the Curonian Lagoon. Therefore the total deposition in the delta can reach about 80% of the brought by the river.

The not implemented project was made to protect these valley areas from the spring floods for intensive farming. It would increase the water contamination in the standing water of south part of the Curonian Lagoon, where and in this time the water is not quite clear. The similar project was realized in Denmark on the delta of the river Skjern. It caused the hard increase of water contamination in the Rinkgobing lagoon, and fishes began to die there. (Ministry..., 1999). Therefore it became necessary to restore the former floodwater flow through the valley. This example shows, how important is to perform the modeling of sediment motion in such cases. In the Nemunas delta only the protecting from summer floods was useful and had been performed. However, for example, in the valley of the river Minija the protection of some valley parts from the spring floods was suitable, as there are large enough not separated areas. In the river deltas such areas are absent mostly (Dolgopolova, 2004).

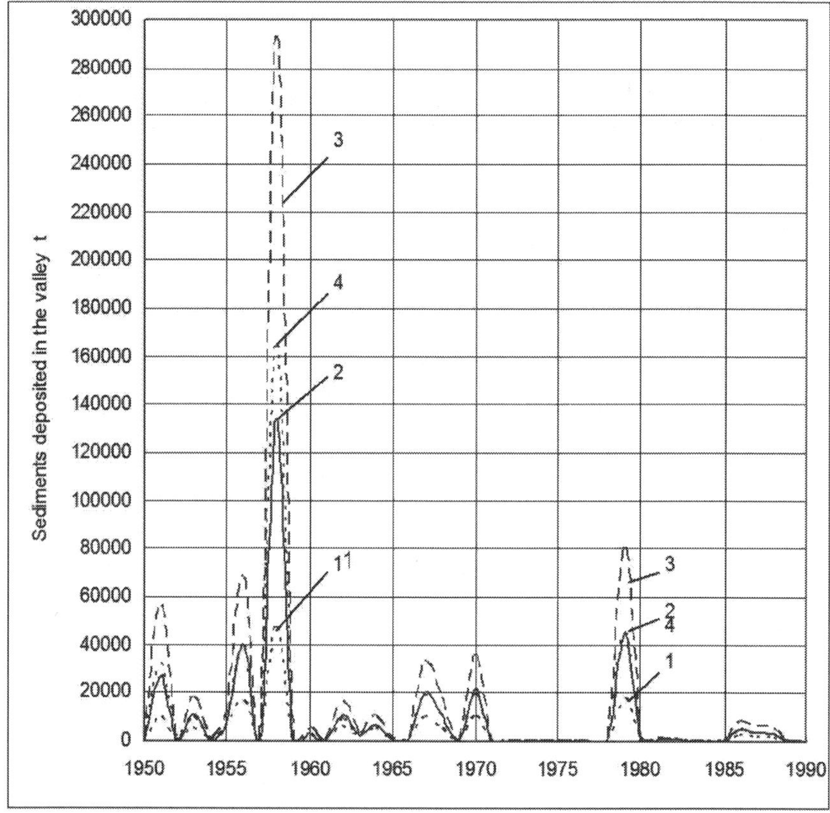

Fig. 4. Sediments deposited in upper strip of valley: 1 – diameter of fraction particles 0.005 mm; 2 – diameter of fraction particles 0.01 mm; 3 – diameter of fraction particles 0.02 mm; 4 – diameter of fraction particles 0.1 mm. The summary amount of sediments settled during 1950–1991 time interval is equal to 1670000 t

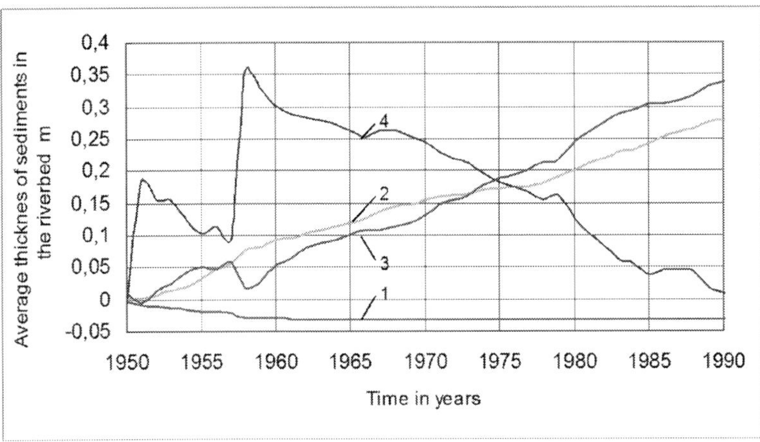

Fig. 5. The thickness of accumulated sediment layer: 1 – in the first segment of riverbed, in the first half of down strip; 2 – in the second segment, in the second half of down strip till the river Gege; 3 - in the third segment, in the first half of upper strip; 4 – in the forth segment, in the upper strip till its top.

3.2 The possibilities to increase the sediment deposition in the valleys

Naturally it would be desirable to find the means to increase the sediment deposition of muddy water still more. Therefore it was investigated the means and factors influencing the sediment deposition in the valleys. (Rimkus & Vaikasas, 2010). These factors are:

1. The increase of water discharge flowing through the valley by deepening and widening of places where the floodwater overflows from riverbed into the valley.
2. The forming before the floods of the best state of the grasses for sediment entrap in the floodplain meadows.
3. The building of way banks across the floodplain
4. The slowing of water stream in the valley by growing of bushes and wood.

For the increase of water discharge overflowing in the upper part of Nemunas delta it was widened and deepened the natural water overflow place existing at the settlement Panemunė (Fig.6). Increase of the conductivity of this overflow was successful, as the water flows further into the wide lake Užlenkė, from which it spreads into the whole valley. Attempt to increase the conductivity of other smaller overflows, for example of Malūnkalnis or Marižiogis, was not successful, as the overflowing water further flows through the narrow beds, which limits their conductivity. Widening only of the inflow from the riverbed is not effective.

The sediment deposition in the valley can be increased also by deepening of wide water overflow place below the railway bridge, however economically effective appeared only the deepening of the overflow at the Panemunė. On the possibility to dig here the channel the attention was turned during the investigations to select the optimal trace for the road round the town Sovietsk. The ground for the road banks could be taken from this channel. Thus it was found by these investigations not only the optimal trace of the road, excluding it's negative influence, it was detected also the possibility to increase and the conductivity of the valley. For the building of this road it would be necessary to dig the channel with the width 120 m and the depth – 2 m.

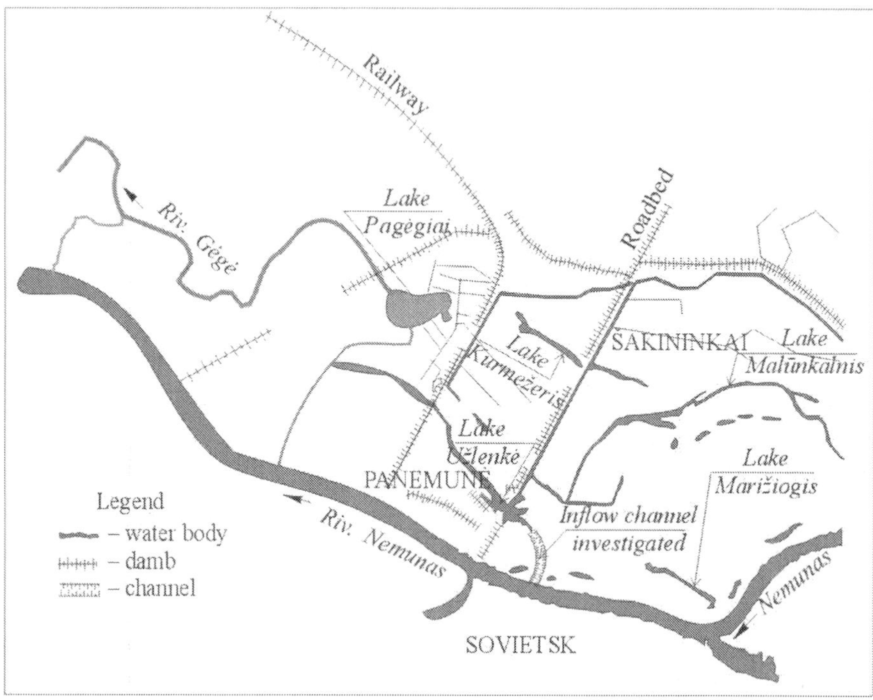

Fig. 6. The main canal and floodplains of the river Nemunas in Panemunė

The built across the Nemunas delta railway and highway decrease the sediment deposition about by 30%, as they are nearly to the main water overflow in valley. However the deepening of the water overflow into the valley at the Panemunė compensates this negative influence and still increases the water discharge flowing in the valley.

The bush and wood growing in the valley decreases the flow velocities there. The decrease of flow velocities in the riverbeds can to increase deposition of suspended sediment. However in the meadows of the floodplains the sediment deposition does not increase with decreasing of flow velocities, as the velocities in flooded valleys are low and the grasses are not laid. Consequently they can to entrap the sediments independently on the little exchange of flow velocities, as their state does not change in these conditions.

In the cases when the thick bushes increase the water level too much, they can pond the water inflow in the valley and to decrease the sediment deposition there.

3.3 Sediment motion dynamics in the River Virvyte

To utilize the renewable energy, hydroelectric power plants (HPPs) are built in rivers. However, their weirs and made ponds affect the natural conditions of the rivers, thus deteriorating the life conditions of water fauna and flora. Ponds have major impact on river hydrology. It was found and in other rivers (Wang et al., 2003, Marčiulioniene et al., 2011). Weirs disrupt the natural motion of sediments and the organic materials accepted with them. The organic silt is mostly retained in the reservoirs, instead of fertilizing the downstream floodplains. In reservoirs, anaerobic processes and algal populations tend to dominate, and eutrophication may occur if there is an excess of nutrients in the water and sediments (WFD, 2007; Povilaitis, 2008). Thus, ecosystems can be influenced to a large extent and the water quality may get worse (Vaikasas & Dumbrauskas, 2010; Rimkus & Vaikasas, 2010; Olli, 2008). Therefore, it is necessary to analyse these processes and reduce their negative influence by choosing optimal methods of hydro-energy employment. This was the second aim of the study.

Further in the chapter, there are presented the results of mathematical-hydraulic modelling of the influence of HPP ponds with various weir heights on the sediment deposition in valleys inundated during the floods. When the weirs are not too high and the riverbed volume is sufficient for arrangement of the ponds, the valleys are inundated only during the floods. In the flooded meadows, the sediments washed from the fields settle steadily. Consequently, the quality of river water is improved greatly because of the settled particles, brought from the adjacent agricultural lands and deposited there, are rich with adsorbed nutrient load (Bakel, 2006). On the contrary, when the weirs are too high, some part of the valley area is always inundated. In this case, the valley area, useful for sedimentation, decreases. Consequently, the water self purification process decreases also, as more significant part of sediments, containing pollutant materials, return from the valley to the river flow. Therefore the too high weirs can worsen the water quality in the downward river reaches.

In deep ponds, there are some other factors that are unfavourable for the formation of water quality. The low stream velocities in large ponds create favourable conditions for algae and other small vegetation to grow. The decayed fine vegetation pollutes the water. The silt sinks on the bottom, although elevated flow velocities could lift it. These velocities increase during the daylong power regulation. With increased turbine discharge, the suspend organic silt mixes with water and passes down. The oxidation of these organic materials decreases the amount of dissolved oxygen over a long interval of the river. This process in the Lithuanian rivers has already been discussed earlier (Vaideliene, 2008; Ždankus, 2008; Ždankus & Sabas, 2005; Zdankus et al., 2008).

Most ponds of the river Virvytė fill up only the riverbed, and only few of them overflow into the valley. The present investigation is based on the modelling of a 12-km interval of the small river Virvytė, where a 11 HPP cascade includes three HPPs: Skleipiai, Kapėnai, and Kairiškiai (Fig.7). In the modelled river interval, the pond of Kapėnai HPP occupies 30 ha of a dammed valley. This decrease the sediment deposition in the valley. The main amount of sediments settled in the valley deposits during the frequent, although not large floods. In a modelled 50-year interval, an area of 60–70 ha at the Kapėnai pond was flooded most frequently. Therefore, it is natural that, in such a long period of time, the sedimentation decreased by half. The average sediment deposition in the valley strips near the riverbed makes 2–3 t/ha/year. The similar quantities were found during the field investigation in the flooded valley of the river Nevėžis (Vaikasas, 2010).

For the ponds in Skleipiai and Kairiškiai, the riverbed is sufficient for their ponds; while in the Kapėnai pond, as it is mentioned above, some part of the pond overflows in the valley. The calculation results show a considerable decrease of the sediment deposition in the valley in the lower region of this pond. The results are plotted in Fig. 8.

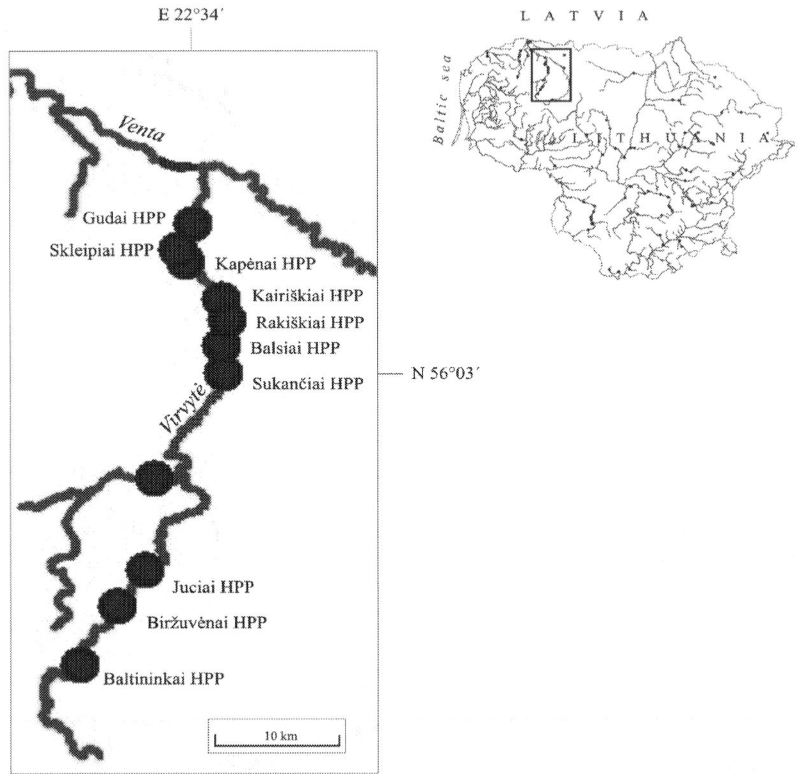

Fig. 7. The HPPs cascade in the river Virvyte

Three groups of curves are plotted in Fig. 8. Thin lines depict the case when the HPP ponds are absent and the sediment deposition is naturally great. Thick lines indicate the case where all ponds are present in the investigated interval and the sediment deposition decreases intensively. During the high floods, the sediment deposition decreases almost by half and, during the small ones, even several times. The dotted lines show the case where only the Kairiškiai pond is equipped. Here, the sediment deposition is practically the same or even somewhat greater compared with the case without HPP weirs. Thus, such a HPP has some positive influence, since even low weirs slightly pond up the flood water levels.

The sediment deposition in the inundated valley increases quickly with rising of water rates, since in this case the sediment discharge also increases, and the decrease of sediment concentration because of their deposition is then compensated. In addition, the area of inundated valley increases with the flood increase. The floodplain area in the region of these three ponds during the flood of a 1% probability covers 400 ha. The deposition of coarse particles increases with discharge growing much more intensively, because the fall velocity of these particles is also much higher.

In Fig. 9 the deposited sediments are expressed as part of the total amount of sediments brought to the river. The deposition is quite intensive. When the water discharge is large, the deposition of fine sediments increases less intensively than in the case of sediments brought by the river. Therefore, the relative deposition of these sediments starts decreasing a little.

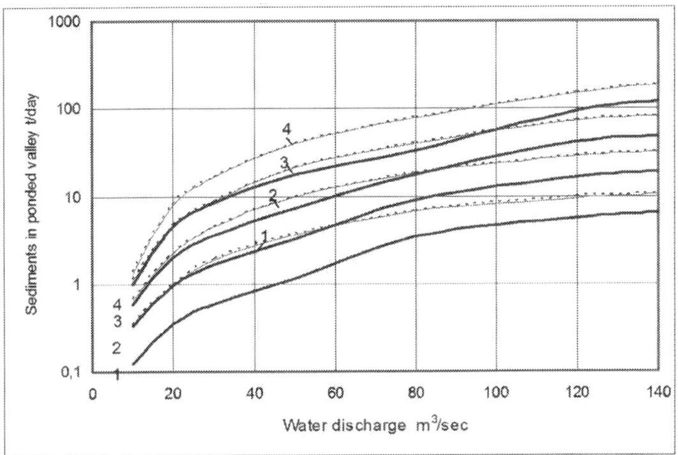

Fig. 8. Amount of deposited sediments in the investigated interval of the river Virvytė valley as a function of flood discharge: thin lines – weirs are absent, thick lines – with all 3 HPP weirs, dotted lines – only with a weir of Kairiškiai HPP. Particle diameters in the sediment fractions are 0.001 (1), 0.002 (2), 0.005 (3), and 0.01 (4) mm.

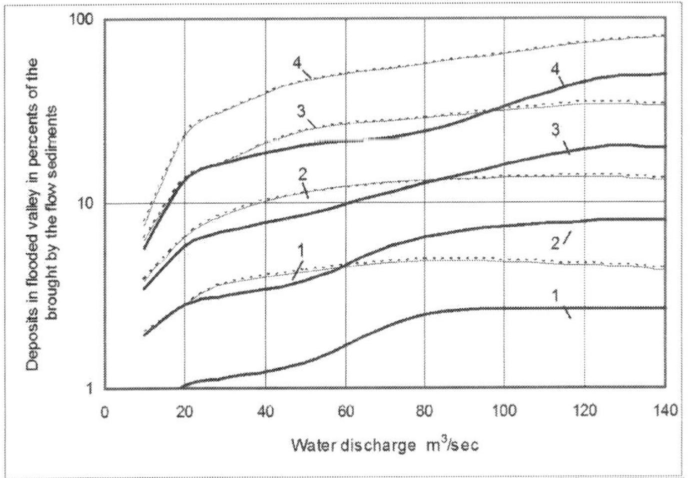

Fig. 9. Intensity of sediment deposition in the inundated valley as part of sediments brought by the river: thin lines – dams are absent, thick lines – with all 3 HPP dams, dotted lines – only with a dam in Kairiškiai HPP. Particle diameters in the sediment fractions were 0.001 (1), 0.002 (2), 0.005 (3), and 0.01 (4) mm.

The arrangement of the Kapėnai pond decreased the sediment deposition in the valley more intensively during the relatively low floods, since this pond took away some of floodplain parts important for the sediment retention, which would be flooded by the low, but more frequent floods.

Fig. 10 depicts the longitudinal profile of the investigated interval of the river Virvytė during the floods with a water discharge of 20, 50, 100, and 150 m³/s. These floods pond the water level below the HPP and somewhat decrease the power of the turbines. In the Kapėnai pond, there is a greater area of cross-sections near the dam, and thus the water level along the river increases less than in the other two ponds. Therefore, the stream velocities in the ponds of Skleipiai and Kairiškiai are higher, and their water levels increase along the flow more intensively. This leads to increase of inundated area of the valley and to increase of sediment deposition. As a result, the floodplain meadows are more fertilized and the water quality of the Virvyte River is improved.

During high floods, the sediment deposition in the meadows of the Virvyte is more intensive; however, such floods are rare. The lower flood discharges occur more frequently and the significant part of sediments is deposited then. To estimate the influence of different floods, the calculations were performed for a long-term period. The results are shown in Figure 11.

The amount of sediments deposited in a one-year period depends on the size of the flood. The most intensive deposition was observed during the large flood with a 1% probability in 1958. Some years, the floods were low and did not overflow in the valley. In this case, the processes of sediment deposition and retention do not proceed. When all three ponds are involved, the sediment retention decreases by about 50% due to the too high weir of Kapėnai HPP.

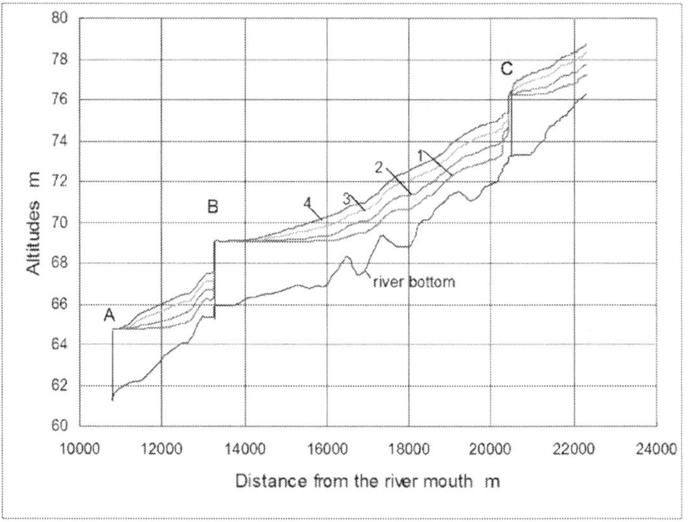

Fig. 10. Longitudinal profile of the investigated interval of the river Virvytė. Water levels of floods with water discharges of 20 (1), 50 (2), 100 (3), and 150 (4) m³/sec. HES dams: A – Skleipiai, B – Kapėnai, and C – Kairiškiai.

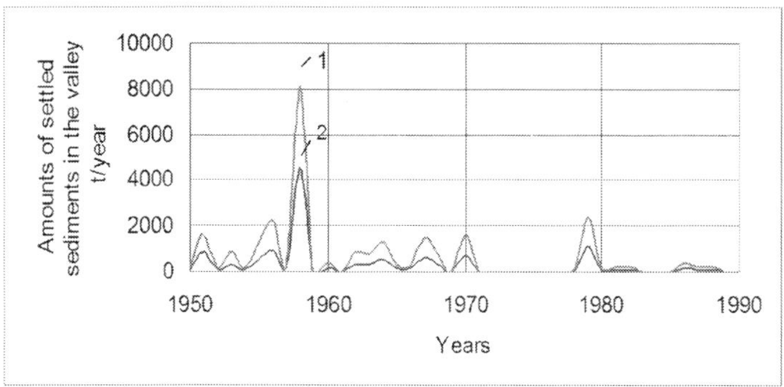

Fig. 11. Amounts of the sediments settled during a many-year period in the interval of Virvyte with the HPP ponds of Skleipiai, Kapėnai, and Kairiškiai: 1 – ponds are absent and 2 – all three ponds are involved.

Thus, the volume of sediments deposited in the floodplain is quite large. Such amounts of sediments per unit floodplain area have been found in our field investigations (Rimkus & Vaikasas, 2010; Rimkus et al., 2007; Lamsodis & Vaikasas, 2005; Vaikasas & Rimkus, 2003). In periodically flooded meadows, the grasses actually entrap the sediments and favour self-cleaning of the rivers. Therefore, a decrease in the grassed areas worsens this natural process. This fact was also confirmed by other researchers (Jankowska-Huflejet, 2006; Habersack et al., 2008; Vaikasas & Dumbrauskas, 2010; Lukianas et al., 2006; Ždankus & Sabas 2005; Zdankus et al., 2008). As one can see, in small rivers, high dams are not desirable from the ecological point of view; they can be important only for energetic purposes, for example, for the daylong regulation of power or for water energy accumulation. One rather large pond at the upper station of a cascade would be enough to successfully regulate the water discharge for the plants in the lower region of the river. It would not worsen the ecological conditions significantly, however such a pond would better supply the local exchange of electricity demands, and all the necessary pike energy could not be transported from the system. This would reduce the energy losses in the electricity supply network.

Ten hydropower stations built on the river Virvytė are grouped in two cascades. In the lower one, the upper station (Sukančiai) has a pond sufficient for a daylong power regulation. However, such a regulation and optimal energy production are impossible now, since these stations are equipped with only one or two large propeller-type turbines (for economic reasons). Usually, they operate at their maximum power, which is much higher than the one ensured by the river; otherwise, their efficiency coefficient would be too low. Having worked down the water level in the pond to a permissible limit, the power units are stopped, and only a sanitary discharge is allowed to pass until the pond is filled again. This situation was discussed by Ždankus and Krakauskas (2000), too. According to the investigations performed in the river Virvytė, such an intensive fluctuation regime is very unfavourable for the environment. The living fish is decreasing in number. Such unfavourable conditions were also found in other rivers (Lopardo & Seone, 2004).

When designing the power stations on the river Virvytė, the necessity to install better turbines has not been considered, because the total power of all river stations was small compared with the power of the whole system. Therefore, the regulation of small HE power stations was not considered to be of significant importance. Moreover, no attention was paid to the ever-increasing possibility of ecological damage.

To improve this situation, at least one Kaplan-type turbine with a wide power regulation should be installed anew or replace the old one at each station. The installation of Kaplan-type turbines would be compensated economically in some period of time, since it would be not necessary to pass a sanitary discharge uselessly, when the utilized volume of pond is refilled. The turbines with a power regulation allow utilizing almost the whole water discharges of the river, except the surplus flowing during the floods. In addition, this makes it possible to produce higher-value energy adapted to the usage exchange. In this way, the ecological and energetic demands will be coordinated. Therefore, it is even supposed that the erection of HPPs in small rivers can be also possible and useful (Bruno, 2009). However, a certain large enough amount of rivers in each region must be left untouched, for preservation of natural environment.

Since the most dams of the river Virvytė are not high, the stream velocities in its ponds during the floods are sufficient for transportation of fine sediments. Therefore, they are not silted by the deposition of silt and clay particles. Only coarse particles brought from the fields settle there. This fact has been proved by analysing the ground samples taken from the bottom in all ponds of HPPs.

4. Conclusions

The flooded meadows in river valleys especially in their deltas are very important for ecology conditions, as the grasses in these valleys entrap the brought by water flow sediments with adsorbed contaminations.

When the protection of river valleys from the spring floods is designing, it is necessary to model the sediment deposition, and to estimate the possible increase of water contamination. In such cases it is necessary the large enough areas not separated by dikes for sediment deposition, which usually are absent in the deltas.

The calculation methods for sediment deposition in flooded grassed valleys have not been performed yet, therefore the new formulae for these calculations are proposed.

When in the rivers the hydro energetic power plants are constructed, the environmental and energetic needs are to be coordinated. When designing power plants on the river Virvyte, this coordination was no performed. Therefore the ecological conditions were heavily worsened.

Water self cleaning in small rivers with HPPs is better, when the weirs are not too high, and the ponds do not inundate in the valleys.

Mostly in small rivers HPP turbines are of cheaper propeller type, which cannot regulate their power, and therefore work periodically with the maximal power. That is very harmful for water flora and fauna. To improve this situation the Kaplan type turbines with power regulation must be installed.

5. References

Bakel, J. (2006). Impact of the WFD on agriculture in Netherlands and possible effect – specific hydrological measures: the Duch approach. *Journal of Water and Land Development*, Vol. 10, pp. 45-53.

Barfield, B.J.; Tollner, E.W. & Hayes, J.C. (1979). Filtration of sediment by simulated vegetation. I. Steady flow with homogeneous sediment. *Trans. ASCE*, Vol. 21, No. 3, pp. 540–545, 548.

Bixio, A. C. & Defina, A. (2004). Mathematical modelling of mean flow and turbulence in vegetated open Channel flows. Proc of *Fifth International Symposium on Ecohydraulics*. September 12-17, 2004, Madrid Spain.pp.765-770.

Bruno, G. S. (2009). Developing European Small Hydro to its full Economic Potencial. *Hydropower & Dams*. Issue 2, pp. 102-108.

Carpena, R.M.; Parsons, I. E. & Giliam, J.W. (1999). Modelling hydrology and sediment transport in vegetative filter strips. *J. Hydrol.*, Vol. 214, pp. 111–129.

Christensen, B.A. (1985). Open channel and sheet flow over flexible roughness. Proceedings 21th Congress IAHR 19–23 August, Melbourne, pp. 462-457.

Christiansen, T. & Wiberg, P. (1997). Sediment deposition on a salt march surface. Available at: http//wsrv.clas.virginia.edu/-tc5e/allsa96.html.

Deletic, A. (2001). Modelling of water and sedimentation transport over grassed areas. *J. Hydrol.*, Vol. 248, pp. 168–182.

Dolgopolova, E.N. (2004). River flow regulation impact upon habitat in nature streams. Proc of *Fifth International Symposium on Ecohydraulics*. September 12-17, 2004, Madrid Spain. pp. 771-776.

Fustec, E.; Mariott, A.; Grillo, X.; Sajus, J. (1991). Nitrate removal by denitrification in alluvial groundwater: role of a former channel. *Journal of Hydrology*, Vol. 123, pp. 337-354.

Habersack, H.; Hofbauer, S.; Hauer, C. (2008). Vegetation impacts on flood flows – evaluation of resistance based on a hydraulic scale model and numerical hydrodynamic modoling. *River Flow* Altinakar, Kokpinar, Audin, Cokgor and Kirkgoz (eds.) ISBN 978-605-60136-1-4, 425-432.

Haycock, N.; Burt, T. (1990). Handling excess nitrates. *Nature*, Vol. 348:29.

Haycock, N.; Burt, T. (1991). The sensitivity of rivers to nitrate leaching; the effectiveness of near–stream land as a nutrient retention zone. In: Allison R.Thomas D.(eds) *Landscape sensitivity*, pp.261-272 John Willey & Sons, Chichester

Jankowska-Huflejet, H. (2006). The function of permanent grasslands in water resourses protection. *Journal of Water and Land Development*, No. 10, pp. 55-65.

Kouwen, N. (1987). Velocity distribution coefficients for grass-lined channels. Discussion. *J. Hydraul. Engng.*, Vol. 113, No. 9, 1221–1224.

Kouwen, N. & Unny, T.E. (1973). Flexible roughness in open channels. *J. Hydraul. Div.*, Vol. 99 (HY5), pp. 713–728.

Jankowska-Huflejet, H. (2006). The function of permanent grasslands in water resources protection. *Journal of Water and Land Development*, No. 10, pp. 55-65.

Lamsodis, R.; Vaikasas, S. (2005). The Potential to Retain Nitrogen in Beaver (*Castor fiber L.*) Ponds and Through Man-Controlled Flooding in the Nemunas River Basin. *Archiv fur Hydrobiologie*, Suppl. – *Large Rivers*, Vol. 15, pp. 227-241.

Large, A.; Petts, G. (1994). Rehabilitation of river margins. In: Calow, P., Petts, G. (eds). *The Rivers Handbook*, Vol. 2, pp. 401-418. Blackwell Scientific Publications, Oxford.

Lopardo, R. A. & Seone, R. (2004). Environmental impact of large and small hydraulic structures. *Fifth Inernational Symposium on Ecohydraulics. Aquatc Habitans. Analysis & Restoration.* Madrid, pp. 867-872.

Lukianas, A.; Vaikasas,S.; Mališauskas, A. P. (2006). *Irrigation and Drainage,* Vol. 55, Issue 2, pp. 145-156. ISSN 1531-0353.

Marčulioniene, D.; Montvydiene, D.; Kazlauskiene, N;, Kesminas, V. (2011). Changes in macrophytes and fish communities in the cooler of Ignalina Power plant (1988-2008). *Journal of Environmental Enginiering and Landscape Management,* Vol. 19, No. 1, pp. 21-33.

Middelkoep, H. & Haselen, O. G. (1999). *Twice a River. Rhine and Meuse in the Netherlands.* RIZA Report.

MIKE 21 (1995). *Sediment Processes. User Guide and Reference Manual.* Danish Hydraulic Institute, Copenhagen.

Ministry of Environmental and Energy of Denmark. (1999) The Skjern river regulation project. Denmark, Copenhagen 33 p.

Olli Gul, (2008). Historic sediment accumulation rates in Karlskarsviken, a bay of Lake Malaren, Sweden. *Hydrology Research,* Vol. 39, No. 2, pp. 123-132.

Pasche, E. & Rouve, G. (1984). Over bank flow with vegetatively roughened flood plains. *J. Hydraul. Engng., ASCE,* Vol. 111, No. 9, pp. 1262–1278.

Povilaitis, A. (2008). Sourse apportionment aand reention of nutrients and organic matter in the Merkys river basin in southern Lithuania. *Journal of Environmental Enginniering and Landscape Management,* Vol. 16, No. 4, pp. 195-20.

Pukštas, R. & Vaikasas, S. (2005). Experimental investigations on the chemical agregate and grain-size composition of spring flood sediment of the river Nevėžis.Transactions *Water Management Engineering,* Vol. 28, No. 48(1), pp. 70-74. ISSN 1392-2335 (in Lithuanian).

Rimkus, A.; Vaikasas, S. (1997). Improvement of Calculations of Suspended Sediment Deposition. *Environmental Reasearch Engineering and Management,* No. 2(5). pp. 29-35.

Rimkus, A.; Vaikasas, S. & Pukštas, R. (2007). Calculation of suspended sediment deposition in grass-covered floodplains. *Nordic Hydrology,* Vol. 38, No 2, pp. 151-163. doi:10.2166/nh.2007.004.

Rimkus, A.; Pukštas, R. & Vaikasas, S. (2004). Investigation on suspended sediment concentration and grain-size composition in the water of river floodplain. In: *Proc. V IAHR International Symposium of Ecohydraulics, 12–17 September, Madrid, Spain* pp. 1416–1422.

Rimkus, A. & Vaikasas, S. (1999). Calculation of the suspended sediment deposition in flooded, overgrown with grass valleys of rivers. In: *Proc. XXVIII IAHR Congress. 22–27 August, Graz, Austria.* CD-ROM,IAHR, 1141-1147.

Rimkus, A.; Vaikasas, S. (2010). Possible ways to improve sediment deposition in the Nemunas delta. *Hydrology Research,* Vol. 41, No. 3-4, pp. 346-354.

Roger, A. F. & Lin, B. (2003). Hydro-environmental modelling of riverine basins using dynamics rate and partitioning coefficients. *Integr. River Basin Mngmnt.,* Vol. 1, No. 1, pp. 81–89.

Soulsby, R.L.; Manning, A.J.; Whitehouse, R.J.S. & Spearman, J.R. (2010). Development of a generic physically – based formula for the settling flux of natural estuarine cohesive sediment. Final Report – summary, H R Wallingford company research project DDY0409 1p.

Temple, D.M. (1986). Velocity distribution coefficients for grass-lined channels. *J. Hydraul. Engng., Vol.* 112, No. 3.

Thornton, C.I.; Abt, S.R. & Clary, W.P. (1997). Vegetation Influence on Small Stream Siltation. *J. Am. Wat. Res. Assoc.,* Vol. 33, No. 6.

Vaideliene, A & Michailow, N. (2008). Dam influence on the river self-purification. In *the 7 International conference "Environmental Engineering"* Vilnius Gediminas Technical University, 2008, pp. 247-251.

Vaikasas, S. (2010). Mathematical modelling of sediment dynamics and their deposition in Lithuanian rivers and their deltas (case studies) *Journal of Environmental Engineering and Landscape Management,* Vol. 18, No. 3, pp. 207-216. ISSN 1648-6897.

Vaikasas, S. & Rimkus, A. (2003). Hydraulic modelling of suspended sediment deposition in an inundated floodplain of the Nemunas Delta. *Nordic Hydrol.,* Vol. 34 , No. 5, pp. 519–530.

Vaikasas, S. & Rimkus, A. (1996). Problems of suspended sediment accumulation in flooded delta of Nemunas River. *Water Management Engineering. Transactions,* Vol. 1, No. 23, pp. 120-137. (in Lithuanian).

Vaikasas, S, Gipiskis, V. & Katutis, K.(1997). The formation of alluvial soils by settling the suspended sedimentsin the flooded Delta Nemunas. *Proc. of the scientific conf.* Vilnius, February 20, 1997 pp. 75-81. ISBN 9986-527-28-7.

Vaikasas, S. & Dumbrauska, A. (2010). Self–purification process and retention of nitrogen in floodplains of River Nemunas *Hydrology Research,* Vol. 41, No. 3-4, pp. 338-345.

Van Rijn, L. C. (1993). Principles of sediment transport in rivers, estuaries and coastal seas. University of Utrecht Netherlands. Aqua Publications

Wang, Z.V., Hus, Vu Y. & Shao, X. (2003). Delta processes and management strategies in China. *Integr. River Basin Mngmnt.,* Vol. 1, No. 2, pp. 173–184.

WFD & Hydropower; (2007). Water Framework Directive & Hydropower. Proceedings of *the Common Implementation Strategy Workshop.* Berlin, 4-5 June 2007: 1-4.

Yurchuk, M. (1987). *Hydraulic Characteristics of Flow in Overgrown Channels* (in Russian). Dissertation. Moskovskij Ingenerno-Stroitelnyj institut, Moscow.

Ždankus, N.; Krakauskas, M. (2000). Influence of sinplification of low powered hydraulic turbine on its efficiency. *Power Egineering,* Vol. 1, pp. 3-9. ISSN 0235-7208. (in Lithuanian).

Ždankus, N. (2008). Interaction of closely located hydropower plants. *Proc. of the 7th International Conference "Environmental Engineering".* May 22-23 2008. pp. 764-768.

Ždankus, N. & Sabas, G. (2005). The influence of anthrepogenic factors to Lituanian rivers flow regime. *Proc. of the 6th International Conference "Environmental Enginering",* May 26-27, 2005, pp. 515-522.

Zdankus, N.; Vaikasas, S.,& Sabas, G. (2008). Impact of a hydropower plant on the downstream reach of a river. *Journal of Environmental Ingenering and Landscape Management*, Vol. 16, No. 3. pp. 128-134.

7

The Vulnerability of the Shingwedzi River, a Non-Perennial River in a Water Stressed Rural Area of the Limpopo Province, South Africa

P. S. O. Fouché[1] and W. Vlok[2]
[1]Department of Zoology, University of Venda
[2]Zoology Department, University of Johannesburg
South Africa

1. Introduction

1.1 Background

The Shingwedzi River drains one of the drier sub-catchments of the South African component of the Limpopo River Catchment which is situated in the north-eastern part of the Limpopo Province of South African (Figure 1). This non-perennial river is a tributary of the Olifants River, or *Rio des Elephantes* as it is known in Mozambique, into which it drains which in turn joins up with the Limpopo River.

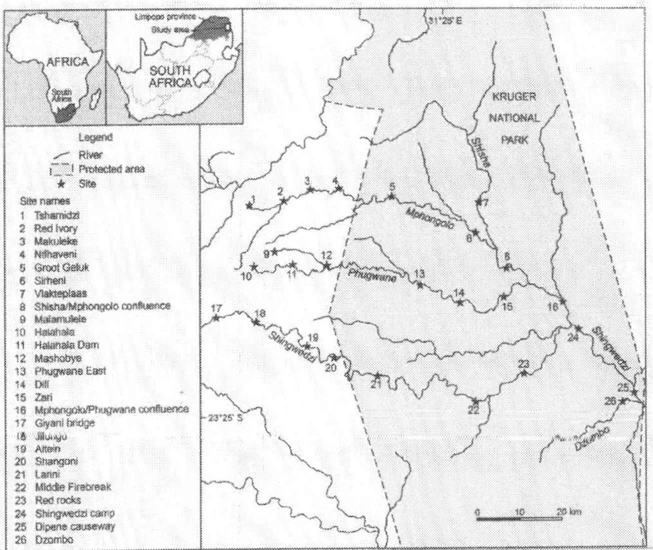

Fig. 1. Map of the study area showing rivers, sampling sites and the boundary of the Kruger National Park, South Africa (Adapted from Fouché and Vlok, 2010).

The Integrated Catchment Management (ICM) approach currently applied in South Africa has resulted in 19 Water Management Areas (WMAs) demarkated within the country's boundaries. The Luvuvhu-Letaba Water Management Area (WMA 02), of which the Shingwedzi River forms part, lies entirely within the Limpopo Province and borders on Zimbabwe and Mozambique (DWAF, 2004a). A unique feature of this WMA is the world renowned nature conservation area, the Kruger National Park (KNP), along its eastern boundary which occupies more than a third of the land area of the WMA. The Shingwedzi River and the two major rivers of the WMA, the Luvuvhu and Groot Letaba Rivers, flow through the KNP into Mozambique. The Luvuvhu River is a direct tributary of the Limpopo River but the Groot Letaba River first flows into the Olifants River before joining the Limpopo River.

The Shingwedzi River, which originates near the town of Malamulele (S 23° 00, 680 E 30° 42,416), has a number of tributaries of which the Mphongolo, Phugwane, Shisha and Dzombo (Figure 1) are the most important. The first two originate outside the KNP whilst the entire catchments of the latter two are within the boundaries of the KNP.

The Shingwedzi River Catchment is relatively small, covering an area of *ca* 5300 km², and the climate is regarded as hot and dry. In addition to low rainfall, which ranges between 400 and 650 mm a⁻¹, it also has a high mean annual evaporation rate of *ca* 1700 mm a⁻¹, resulting in a low annual run-off (Table 1) which, typical to South Africa, displays a historically high degree of variability as is illustrated by the data records of the Shisha River, one of the major Shingwedzi River tributaries (Figure 2). The catchment forms part of the summer rainfall region of South Africa and it typified by dry winters. Average temperatures range between 2,4°C in winter to 40,8°C in summer and the area is mostly frost free.

Sub-catchment	Natural MAR	Ecological Reserve
Luvuvhu	520	105
Shingwedzi	90	14
Groot Letaba	382	72
Klein Letaba	151	20
Total	**1053**	**241**

Table 1. The mean annual runoff (MAR) and recommended Ecological Reserve (million m³ a⁻¹) of the sub-catchments in the Luvuvhu-Letaba Water Management Area in South Africa (Adapted from DWAF, 2004a, 2004b).

The topography of sub-catchment is characterised by plains with a low to moderate relief in the east, giving rise to open hills, while low mountains with high relief are present towards the west (Midgley *et al.*, 1994). Based on the geology, three regions can be distinguished. The first region consists of the upper reach of the Shingwedzi River and actually includes only a small section of river. The second region lies mostly to the west of the KNP boundary and the third region contains the lower reaches of the Shingwedzi River and is predominantly inside the KNP. The first region consists mainly of basalts of the Letaba Formation (Lebombo Group, Karoo Supergroup). The second region consists of potassium-poor quartz-feldspar of the Goudplaats Gneiss Basement, with some Letaba basalts (Karoo

Supergroup). The third region is made up mainly of Goudplaats Gneiss and Makhuttswi Gneiss with a small contribution from ultramafic metavolcanics and metasediments of the Giyani Greenstone Belt (Swazian Erathem). The erodibilty of soils throughout the whole sub-catchment is high and according to Midgley *et al.* (1994) soil depths are moderate to deep in the east, where acid and intermediate intrusives occur and sandy loam soils dominate whereas it is moderate to deep in the west, with its basic/mafic lavas, where the soils are clayey. In the north and south of the sub-catchment small patches of intercalated assemblages of compact sedimentary and extrusive rocks appear.

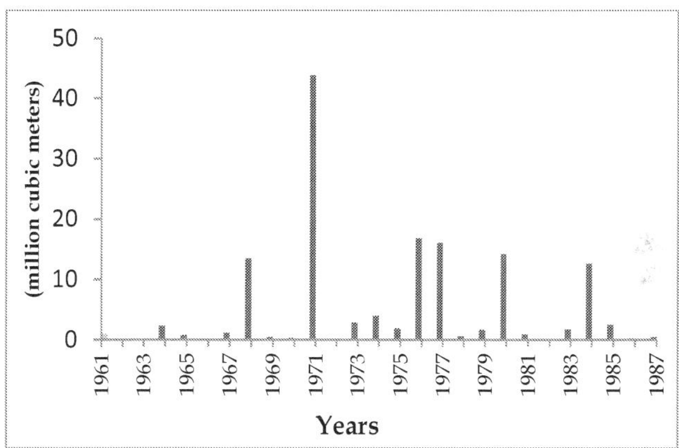

Fig. 2. The Mean Annual Runoff recorded between 1961 and 1987 on the Shisha River, a tributary of the Shingwedzi River, South Africa (Midgley *et al.*, 1994).

The vegetation in the sub-catchment is diverse and Mucina & Rutherford (2006) recognise the distinct vegetation units listed in Table 2 in the sub-catchment. Based on the vegetation and climate the whole sub-catchment is regarded to fall within one region which can generally be referred to as Lowveld or which is referred to by Kleynhans *et al.*, (2007b) as the Bushveld Basin Ecoregion.

Vegetation Units	Dominant woody vegetation
Cathedral Mopane Bushveld	*Colophospermum mopane*
Mopane Basalt Shrubland	*Combretum* sp / *Colophospermum mopane*
Tsende Mopaneveld	*Colophospermum mopane* / *Acacia nigrescens*/*Combretum sp*
Lowveld Rugged Mopaneveld	*Colophospermum mopane*
Mopane Gabbro Shrubland	*Colophospermum mopane*
Makuleke Sandy Bushveld	*Pterocarpus rotundifolius* / *Combretum collinum*
Nwambyia-Pumbe Sandy Bushveld	*Combretum apiculatum*
Granite Lowveld	Mixed *Combretum* / *Terminalia sericea*

Table 2. The vegetation units recognised in the Shingwedzi catchment (Mucina & Rutherford, 2006).

With regard to development the sub-catchment differs from the rest of the WMA where economic activity is characterized by irrigation, afforestation, tourism and commercial and informal farming. Most of the areas outside the KNP are dominated by rural settlements, informal farming and very little industrial development. Midgley *et al.* (1984) reported that the land listed for formal irrigation covers an estimated 2,9 km². Small scale mining operations, of which the majority is defunct, are dotted through the landscape. Under natural conditions the quality of the surface water is good but water of high mineral content occurs in some of the drier parts (DWAF 2004b). Bacteriological pollution of the surface water is a result of run-off from cattle pens and rural villages with insufficient sanitation infrastructure and services. The water resources within the WMA are nearly fully utilized resulting in limited options for further resource development. Reconciliation of water availability and requirements, based on data of the year 2005 (Table 3) show that requirements, with the exception of the Luvuvhu and Shingwedzi River sub-catchments, exceeded the available resources (DWAF 2004a, 2004b). The zero balance in the Shingwedzi sub-catchment is reason for concern because not only is water use regarded as "neglible" (DWAF 2004b) but the available amount of water is based primarily on groundwater resources. The concern is strengthened by the statement that the "over-exploitation of groundwater in the sub-catchment is not sustainable whilst there is insufficient knowledge on the long-term sustainable yield from groundwater and the interdependencies with surface water" (DWAF 2004b). It should further be borne in mind that the reconcilliation, and in particular the requirements, was based on an estimated population of 135 000 people (DWAF 2004a) in the sub-catchment at that time. If the trend in population growth in this region is similar to the rest of South Africa it would be correct to surmise that the Shingwedzi River and its tributaries could come under severe pressure as the need for more water will increase in the near future. Based on the above it is important to take cognisance of the fact that if the increased water requirements for human needs are met using surface water it could become difficult to meet the ecological reserve (Table 1). These deficits could lead to a failure in supplying an adequate amount of water of appropriate quality to address the ecological requirements of the KNP and the honouring of international obligations to Mozambique. Both of these aspects are key considerations in the South African National Water Resources Strategy (NWRS) (DWAF 2004b).

Sub-catchment	Local yield	Local requirement	Balance
Luvuvhu	147	98	**49**
Shingwedzi	3	3	0
Groot Letaba	159	196	(37)
Klein Letaba	32	37	(5)

Table 3. Reconciliation of the water availability and requirements for the year 2005 (million m³ a⁻¹) of selected sub-catchments in the Luvuvhu-Letaba Water Management Area, South Africa. Figures in brackets are negative (Adapted from DWAF 2004b).

Although it is often stated that no major dams occur in the Shingwedzi River system (DWAF, 2004b; Midgley *et al*, 1994) it should be noted that the Makuleke Dam (S 22° 52,046 E 30° 54,377) has been constructed in the Mphongolo River, while there are dams in the upper catchment, near the town of Malamulele, and in main the stem of Shingwedzi River.

In the latter case there is the Kanniedood Dam near the Shingwedzi Rest Camp and the Sirheni Dam near a rest camp with the same name, in the KNP. Although no or little water extraction is done at these impoundments, the impact that they have on the flow regime and connectivity of the system can not be ignored.

As far as the water quality perspectives are concerned it is stated (DWAF 2004b) that phosphate, which result from diffuse sources such as the run-off from agriculture and informal domestic wastewater, is regarded as the only parameter that is adversely affected by activities in the sub-catchment. In addition point sources of pollution, that include mining effluents and treated sewage effluents, could have negative impacts on other water quality parameters.

According to Kleynhans and Louw (2007) EcoClassification is a term used in South Africa for "the ecological classification process for river ecosystems and refers to the determination and categorisation of the Present Ecological State (PES) as well as the health or integrity of various biophysical attributes of rivers relative to the natural or close to the natural reference condition". The purpose of the EcoClassification process is to gain insight and understanding into the causes and sources of the deviation of the PES of biophysical attributes from the reference condition. In South Africa the EcoClassification process forms an integral part of a number of methods such as Ecological Reserve and Environmental Flow Requirement determinations. The methodology is also applied in the national River Health Programme (RHP) where it is used to establish biological response as an indicator of ecosystem health and only assesses cause and effect relationships in general terms.

A number of indices that form part of a suite of tools, originating to a large extent from those developed for use in the RHP, has been adapted for use in the EcoClassification process where it is used to determine the Ecological Category (EC) of a river or reach of a river. The indices, each *inter alia* characterised by a strict protocol, were developed following a Multi Criteria Decision Making Approach (MCDA) (Kleynhans & Louw, 2007) and include the driver asssessment indices (the Hydrological Driver Asessment Index, the Geomorphology Driver Assessment Index and the Physico-chemical Driver Assessment Index) and the biotic response indices (the Fish Response Assessment Index, the Macro Invertebrate Response Assessment Index and the Riparian Vegetation Response Assessment Index).

Kleynhans (2007) refers to the Fish Response Assessment Index (FRAI) as "an assessment index based on the environmental intolerances and preferences of the reference fish assemblage and the response of the constituent species of the assemblage to particular groups of environmental determinants or drivers". The Vegetation Response Assessment Index (VEGRAI) is "designed for the qualitative assessment of the response of riparian vegetation to impacts in such a way that that qualitative ratings translate into quantitative and defensible results "(Kleynhans *et al.*, 2007b). Thirion (2007) describes the Macro Invertebrate Response Assessment Index (MIRAI) as an index that "is used to determine the invertebrate EC by integrating the ecological requirements of the invertebrate taxa in a community or assemblage and their response to modified habitat conditions". A second macro-invertebrate based index, the South African Scoring System (SASS) which in actual fact is the forerunner of MIRAI, was developed as an indicator of water quality. According to Thirion (2007) it has become clear that SASS also gives a general indication of the present state of the invertebrate community but it does not have a particularly strong cause-effect basis.

Physico-chemical monitoring has traditionally been the backbone of water quality monitoring in many countries including South Africa (DWAF, 1986). These results are however representative of conditions at the instant of sampling and do not provide information about the effects of these changes on biological communities and in particular do not provide an insight into historic conditions. The use of diatoms to assess conditions in the aquatic environment has a long history and diatom indices have been developed for various impacts such as salinity, pH, oxygen requirement, nitrogen metabolism, the trophic state which includes inorganic nitrogen and phosphorus concentrations, saprobity or organic enrichment reflected by biological oxygen demand and desiccation (Fouché & Vlok 2009). These indices are based on the fact that diatoms are sensitive to, and appear to have a consistent tolerance to a wide range of environmental parameters. Diatoms have extensively been studied in South African river systems (Schoeman, 1982; Passy *et al.*, 1997; Hardy *et al.*, 2004; Taylor *et al.*, 2007a) and efforts have been made to relate diatom to water quality (Archibald, 1972). Benthic diatom assemblages, being sessile, are exposed to water quality changes at a site over a period of time (Breen, 1998). Initially the approach used for prediction of ecological conditions using diatoms was focused on the abundance of ecologically known taxa but this was refined to indicator-based environmental predictions (Watanabe *et al.*, 1988) where diatom taxa that are tolerant to pollution will always be present in high numbers (Birks *et al.*, 1990). As a result an "indicator value" is always included in an index in order to provide greater weight to those taxa which are good indicators of particular environment conditions.

1.2 Rationale for the study

Although it feeds into the KNP, which is often regarded as the flagship conservation area of South Africa, the Shingwedzi sub-catchment has been neglegted probably due to the fact that "for practical purposes no sustainable yield is derived from surface flow and water use is neglible"(DWAF 2004b). Even though the ecosystem health of a substantive number in South African Rivers have been determined in recent years, the Shingwedzi River has not been part of the effort of the national River Health Programme (RHP) (Strydom *et al.*, 2006). To an extent this is contradictory to the importance of a river that lies within a water-scarce area. In addition a survey of the literature (Gaigher, 1969; Pienaar, 1978; Russell, 1997; Olivier, 2003) showed that a paucity of data existed with regard to fish and this lack of knowledge was in particular severe with regard to areas outside the borders of the KNP.

As part of the project it was hypothesised that the ecological status of the Shingwedzi River is under pressure due to impacts on the drivers and that this is reflected by negative changes in the biological responses which include the fish and riparian vegetation.

2. Materials and methods

2.1 Site selection and surveys

Prospective sites, that were regarded as representative of the different river reaches, were selected using 1:50 000 maps of the area during a desktop study. This was followed by an aerial survey designed not only to investigate the suitability of the sites for monitoring but to detect and locate point and non-point sources of pollution and other anthropogenic impacts on the river. The selection of the sites were then adjusted to include the downstream effects of potential pollution sources such as de-commissioned mines, villages, sewage treatment works and areas of large-scale farming activities. Following the aerial survey all the identified sites were "ground-truthed"during a pilot survey after which the best suited sites were selected for the survey. Each site was surveyed twice, with one survey during summer and the other during winter which also in effect represents the high flow and low flow seasons respectively.

At each site the protocol of the Geomorphological Index (Rowntree *et al.,* 2000) was followed and aspects such as the bank stability, the habitat diversity, erosion and the habitat cover assessed. The gradient of a river segment between consecutive sites was calculated using the recorded altitude of the sites and the distance between the sites. These gradients were used to deduct the zone class and longitudinal zones for each segment and then a "long profile" (Rowntree & Wadeson, 1999) for the Mphongolo, Phugwane and Shingwedzi rivers were drawn.

2.2 Impacts

During the aerial surveys impacts were identified and their location recorded using a handheld GPS. Limited video footage and photographs were captured to illustrate sources of pollution, habitat modifications and other land uses that had a negative impact on the environment. These impacts were then verified and rated during the surveys. The impacts at the survey sites were also noted and recorded. While traveling between sites during the surveys additional impacts were noted and recorded.

2.3 *In situ* determinations and water samples

At each site the pH, dissolved oxygen, electrical conductivity, total dissolved substances and temperature was determined *in situ* with handheld Eutech meters. Sub-surface water samples were collected in acid treated bottles, placed on ice and transported to the laboratory for the determination of nutrient content and total suspended solids.

2.4 Biological indices

The status of the riparian vegetation was determined by applying the protocol of the Vegetation Response Assessment Index (VEGRAI) (Kleynhans *et al.,* 2007b) and selected aspects of the Riparian Vegetation Index (Kemper, 2001). At all the sites a minimum length of 100m of the riparian zone on both the right and left hand river banks were surveyed. The decision whether species in the woody component could be regarded as riparian was based on the findings of van Wyk & van Wyk (1997), Grant and Thomas (2000, 2001) and Schmidt *et al.* (2007).

At each site, where water was present, the relevant macro-invertebrate biotopes were identified and the macro-invertebrates sampled at hand of the protocol used for the South African Scoring System (SASS ver.5) (Dickens & Graham, 2005) and MIRAI (Thirion, 2007). According to the SASS 5 scoring system a sensitivity value with regard of their tolerance towards organic pollution in the water is allocated of the families. These sensitivity scores range between 15 for the sensitive group which includes the Oligoneuridae and Prosopistomatidae, to 1 for the least sensitive group which includes the Muscidae and Culicidae. In this study an arbitrary cut-off point at 8 was taken as the separation between sensitive and non-sensitive. Due to the fact that that SASS was developed as an indicator of water quality it was decided calculate the SASS scores which could then be related to the recorded *in situ* and determined water quality parameters.

With regard to the Fish Response Assessment Index (FRAI) each site was photographed, the biotopes or flow-depth classes (Kleynhans, 2007) identified, demarcated and a sketch map drawn. Overhanging vegetation and undercut banks were identified, their extent estimated and scored. In fast-deep and fast-shallow biotopes the fish were electro-narcotized and collected with scoop nets. Where possible, due to the presence of crocodiles and snags, a small seine net was used in the slow-deep and slow–shallow biotopes. Where this was not possible fish were sampled with a cast net. In small pools, backwaters and in particular where sampling had to be done under and amongst vegetation a pole-seine net was used. All the specimens collected were identified using the key from Skelton (2001). The data collected was used to populate the FRAI model and calculate the scores. To calculate the FRAI scores it is imperative that the reference state (RS) of the river be established (Kleynhans, 2007; Kleynhans & Louw, 2007). In this study the RS was established at the hand of available historic data or derived from similar neighbouring rivers in the same ecoregion (Kleynhans, 2007). The Frequency of Occurrence (FROC) ratings of the expected fish species was obtained from Kleynhans *et al.* (2007a).

Diatom sampling, at the sites where water occurred, and preservation was done at hand of the methods of Taylor *et al.* (2007a). Following sample preparation diatoms were microscopically identified using the key provided by Taylor *et al.* (2007b) and the data was then used to populate the following indices: the Specific Pollution Sensitivity Index or SPI, the Biological Diatom Index or BDI (Lenoir & Coste, 1996) and the Trophic Diatom Index or TDI (Kelly & Whitton, 1995). In addition the Water Quality Index (WQI) (Bate *et al.*, 2004), which is a scale to estimate water quality using measured water quality parameters, was calculated.

3. Results

3.1 The selected sites and sampling frequency

A total of twenty six sites were selected in the river system both in- and outside the boundaries of the KNP (Figures 1 and Table 4). The site in the Tshamidzi River, a tributary of the Mphongolo River, was selected as the highest upstream point in the sub-catchment of the respective river systems and represented what could be regarded as a "least impacted" site. The sites in the Shisha and Dzombo rivers, sites 7 and 26 respectively, were selected to represent two smaller tributaries that originate within the boundaries of the KNP and could therefore be regarded as the least impacted rivers within the Shingwedzi River system. Figure 3 shows that although the Shingwedzi sub-catchment falls within one ecoregion (Ecoregion 3 or Bushveld Basin) three level 2 subregions (3.02, 3.03 and 3.05) could be identified and Table 5 shows the distribution of the sites within these ecoregions and supplies a brief description of each sub-region. While the sites inside the KNP were surveyed in June 2007 (winter, low flow) and March 2008 (summer, high flow) the sites outside were surveyed in May 2007 (winter, low flow) and February 2008 (summer, high flow).

3.2 River zonation and long profiles

The long profiles of the three rivers (Figure 6), created by plotting the altitudes of the sites, are all very similar and no sharp decrease in altitude in any of the rivers was observed. Although the altitudes at which all three rivers originate are similar the origin of the Mphongolo at Tshamidzi is at the lowest altitude.

The calculated gradients between sites and the resultant zone class according to Rowntree & Wadeson (1999) for the three rivers are shown in Table 6. Based on these classes the whole of the Mphongolo River and the whole of the Shingwedzi River, with the exception of the section between Jilongo and Altein, can be classified as "lower foothill" (Table 7). Although the gradient of the section between Jilongo and Altein causes it to be classified as a "lowland river" it should be noted that the calculated gradient is very close to the lower end of the gradient of "lower foothill". With the exception of the section between Halahala and Mashobye, which is classified as "upper foothill" the Phugwane River is classified as "lower foothill". Lower foothill zones are described by Rowntree & Wadeson (1999) as zones i) with a mixed bed alluvial channel dominated by sand and gravel, ii) where locally there may be bedrock controls,   iii) where reach types typically include pool-riffle or pool-rapid and iv) where sand bars are common in pools. According to these authors "pools form a significantly greater component than rapids or riffles and flood plains are often present in these river zones".

Site no	Site name	River	Coordinates	Site no	Site name	River	Coordinates
1	Tshamidzi	Mphongolo (Tshamidzi)	22° 53.565 30° 46.180	14	Dili	Phugwane	23° 01.905 31° 07.851
2	Red Ivory	Mphongolo	22° 52.388 30° 49.065	15	Zari	Phugwane	23° 04.115 31° 13.517
3	Makuleke	Mphongolo	22° 51.699 30° 55.629	16	Mphongolo Phugwane confluence	Phugwane	23° 01.146 31° 19.283
4	Ntlhaveni	Mphongolo	22° 52.846 30° 57.489	17	Giyani Bridge	Shingwedzi	23° 04.383 30° 40.537
5	Groot Geluk	Mphongolo	22° 52.783 31° 02.613	18	Jilongo	Shingwedzi	23° 07.455 30° 50.764
6	Sirheni	Mphongolo	22° 57.859 31° 15.098	19	Altein	Shingwedzi	23° 08.234 30° 54.071
7	Vlakteplaas	Shisha	22° 52.165 31° 13.450	20	Shangoni	Shingwedzi	23° 10.623 30° 56.234
8	Mphongolo/ Shisha confluence	Mphongolo	22° 58.344 31° 15.191	21	Larini	Shingwedzi	23° 12.610 31° 05.460
9	Malamulele	Phugwane	23° 00.680 30° 42.416	22	Middle firebreak	Shingwedzi	23° 12.538 31° 13.935
10	Halahala	Phugwane	23° 03.538 30° 47.021	23	Red Rocks	Shingwedzi	23° 10.321 31° 18.419
11	Halahala Dam	Phugwane	23° 03.464 30° 47.474	24	Shingwedzi	Shingwedzi	23° 06.433 31° 26.255
12	Mashobye	Phugwane	22° 59.965 30° 52.849	25	Dipene	Shingwedzi	23° 23.191 31° 33.224
13	Phugwane East	Phugwane	22° 59.996 31° 01.927	26	Dzombo	Dzombo	23° 13.306 31° 33.110

Table 4. The site numbers, names and position of the sites surveyed in the Shingwedzi River sub-catchment, South Africa.

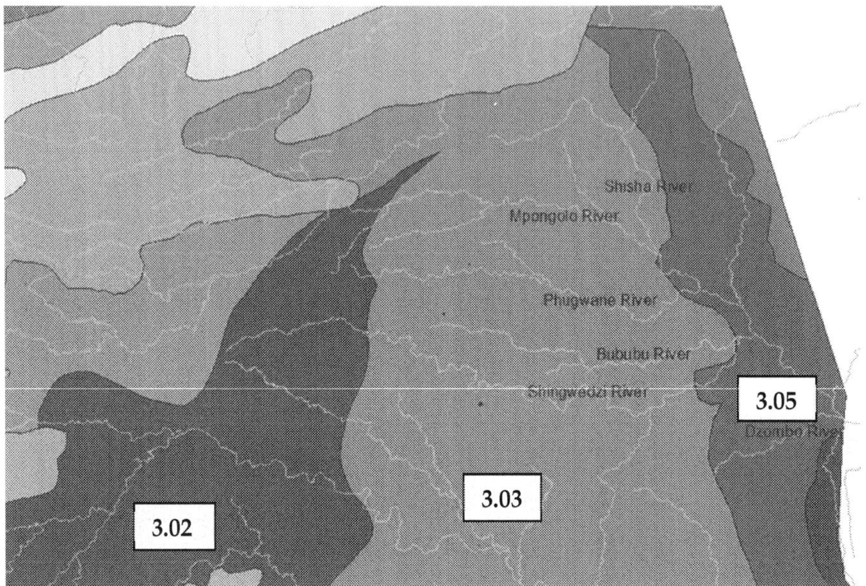

Fig. 3. Map of the study area and the location within the three level 2 Ecoregions of the rivers in the Shingwedzi River sub-catchment, South Africa (Adapted from Kleynhans & Hill, 1999).

Level 2 ecoregion	Ecoregion description	River/Tributary	Sites	Reaches used for FRAI calcultions
3.02	Mixed Lowveld bushveld	Mphongolo and Tshamidzi	1,2	I
3.03	Mopane Bushveld	Mphongolo and Shisha	3, 4,5, 6, 7 and 8	II
3.03	Mopane Bushveld	Phugwane	9,10, 11,12, 13, 14, 15 and 16	III
3.03	Mopane Bushveld	Shingwedzi and Shisha	17, 18, 19, 20, 21, 22 and 23,	IV
3.05	Lebombo Arid Mountain Bushveld	Shingwedzi	24, 25 and 26	V

Table 5. The distribution of the sampling sites within the level 2 ecoregions (Kleynhans *et al.*, 2007b) identified in the Shingwedzi River sub-catchment, South Africa.

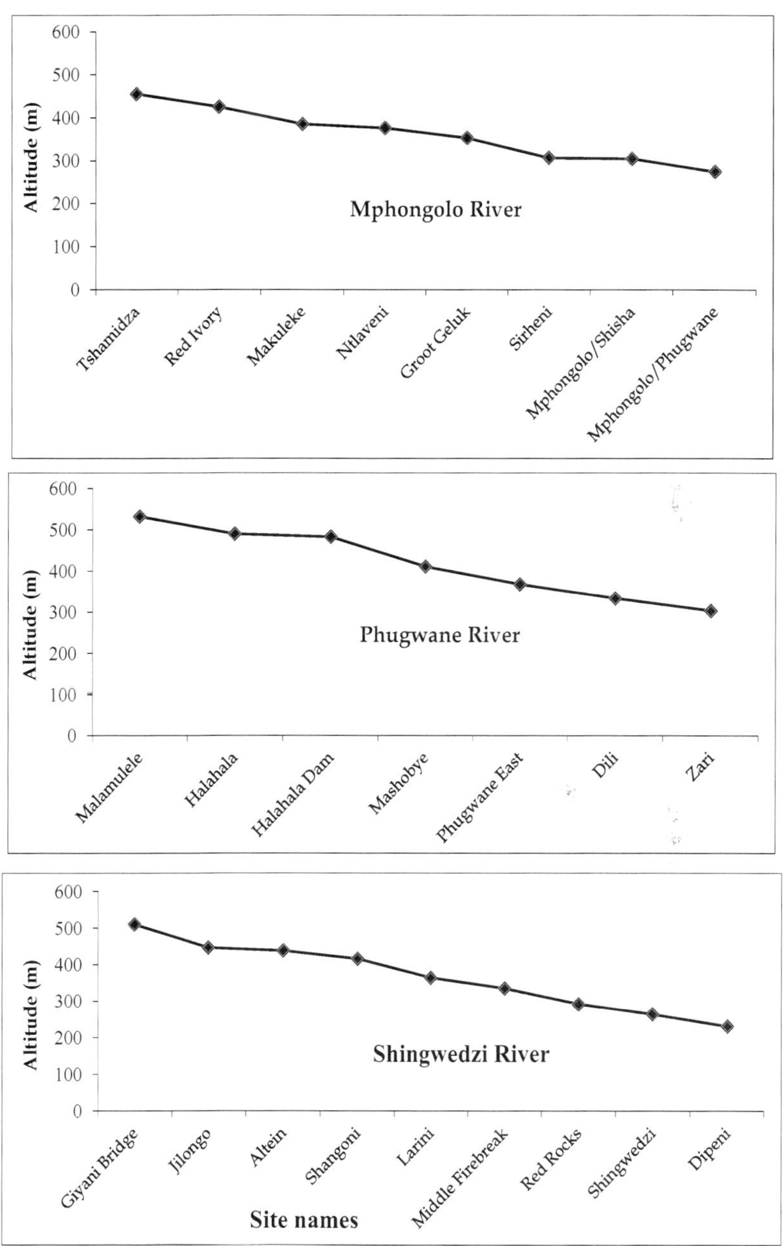

Fig. 4. Long proflies of the Mphongolo, Phugwane and Shingwedzi rivers in the Shingwedzi
River sub-catchment, South Africa.

Section	Distance between sites (m)	Calculated gradient	Zone class
Mphongolo River			
Tshamidzi - Red Ivory	6700	0.004478	E
Red Ivory – Makuleke	14100	0.002837	E
Makuleke – Ntlaveni	5500	0.001636	E
Ntlaveni – Groot Geluk	13400	0.001716	E
Groot Geluk – Sirheni	29100	0.001581	E
Sirheni - Mphongolo/Shisha	15400	0.001429	E
Mphongolo/Shisha – Mphongolo/Phugwane	12200	0.002459	E
Phugwane River			
Malamulele – Phugwane confluence (above Mashobye)	25500	0.004745	E
Halahala - Halahala Dam	1000	0.007	D
Halahala Dam – Mashobye	14300	0.005035	D
Mashobye - Phugwane East	22100	0.001946	E
Phugwane East – Dili	14100	0.002411	E
Dili – Zari	13500	0.002222	E
Shingwedzi River			
Giyani Bridge – Jilongo	27700	0.00231	E
Jilongo – Altein	9900	0.000808	F
Altein – Shangoni	10400	0.002115	E
Shangoni – Larini	27800	0.001871	E
Larini - Middle Firebreak	25100	0.001155	E
Middle Firebreak - Red Rocks	11000	0.003909	E
Red Rocks – Shingwedzi	22200	0.001261	E
Shingwedzi – Dipeni	20400	0.001618	E

Table 6. Calculated gradients and zone classes for the sectors in the Shingwedzi River sub-catchment, South Africa.

Longitudinal zone	Macro-reach characteristics		
	Valley form	Gradient class	Zone class
• Source zone	V10	not specified	S
• Mountain headwater stream	V1, V3	> 0.1	A
• Mountain stream	V1, V3	0.04 - 0.99	B
• Transitional	V2, V3, V4, V6	0.02 - 0.039	C
• Upper foothills	V4, V6	0.005 - 0.019	D
• Lower foothills	V8, V10	0.001 - 0.005	E
• Lowland river	V4, V8, V10	0.0001- 0.001	F

Table 7. The geomorphologic zonation of river channels (Rowntree & Wadeson, 1999).

3.3 Impacts

To simplify the discussion of the impacts the rivers were divided into four sections. Section 1 consists of the Mphongolo River up to the confluence with the Phugwane River and sites 1 to 8 are included, Section 2 is the Phugwane River up to its confluence with the Mphongolo River and includes sites 9 to 15, section 3 is the Shingwedzi River upstream of the Mphongolo River confluence and includes sites 17 to 23 while the Shingwedzi River downstream of the Mphongolo confluence to eastern boundary of the KNP, which includes sites 24 to 26 make up section 4. Only the impacts that are regarded as those that will have the most influence in each section are discussed.

In section 1, in the area outside the KNP, the impacts that increased erosion, such as vegetation clearing, grazing and road crossings were rated as "moderate". While the same rating applied to aspects such as sewage and sand mining, the extensive impacts caused by the irrigation scheme downstream of Site 3, where water abstraction for the 20 centre pivots occurred were rated as "high". In addition the effect of return flow containing pesticides and liquid fertilizers from this irrigation scheme should be noted. In section 2 the general impacts and their rating were similar to that in section 1. What should however be noted is the severe impact of the abandoned mining operation situated between sites 9 and 12 where the dysfunctional slimes dams are severely degraded and runoff could be entering the system after rain events. In the same area the water treatment works and sewage treatment works are not well maintained and there is evidence that the return flows from these facilities are polluting the system. The impact of both the mine and the sewage works was rated as "extremely high". Section 3 is probably the most impacted of all sections. The impacts included solid waste dumping and extensive grazing upstream of Site 17, water extraction at Site 17, brickworks that mine sand upstream of Site 18 and an abandoned mine and sewage return flow from the town of Altein, upstream of Site 19. All of these impacts were rated as "high". Within the KNP the impacts were rated as "low" and consisted mostly of road crossings leading to erosion in section 4. Cognisance should be taken of the return flow from Shingwedzi rest camp and associated sewage works which could lead to nutrient enrichment if not properly maintained. Other impacts are the impoundments in the river which leads to a breakdown in river connectivity. These impoundments include the dams upstream of Sirheni (Site 6) and Malamulele (Site 17), the large lake upstream of Makuleke (Site 3) and the Kanniedood Dam between sites 24 and 25. In addittion the low water bridges at sites 24 and 25 increases the impact on the river through erosion and flow modification.

3.4 The water quality parameters

The recorded winter and summer *in situ* water quality results are shown Tables 9 and 10. In order to relate to their possible effects these results should be read in conjunction with the concept of Thresholds of Potential Concern (TPCs) and the values set by the South African National Parks Board (SANparks) for the KNP. Fouché & Vlok (2010) pointed out that "as part of their management strategy the KNP Management realised the importance and applicability of TPCs and acknowledged that monitoring programmes and associated management interventions are interlinked". In order to achieve their short and long term objectives the KNP then set end-points or TPCs (Table 8) which, when clearly articulated, would contribute towards the strong goal-setting or objectives hierarchy approach (KNP, 2009).

Parameter	Set TPC value
In situ **physico-chemical aspects**	
pH	6,5 – 8,5
Electrical conductivity (µScm⁻¹)	700
Total dissolved substances (mgl⁻¹)	450
Chemical constituents (mgl⁻¹)	
Phosphorous	0,1
Nitrites and nitrates	6
Ammonium	15
Calcium	32
Magnesium	30
Potassium	50
Silicon	18

Table 8. Thresholds of Potential Concern (TPC) values for the Shingwedzi River set by SANParks (KNP, 2009).

During the winter survey the pH values at sites 18, 19, 20 and 22 (Table 9) exceeded the set TPC value (Table 8) with the highest value of 9,09 measured at Site 18. Similar results were recorded during the summer survey (Table 10) with the TPC exceeded at seven sites and the highest value of 9,13 recorded at Site 15. A similar trend where the TPC was exceeded was observed with the electrical conductivity with the highest value of 2590 µScm⁻¹ recorded at Site 14 during winter. In summer, when flow volumes increased, the maximum values were lower than during winter and the highest value recorded was the 1437 µScm⁻¹ recorded at Site 7. Other high values of 1278 and 1029 µScm⁻¹ were recorded at Site 26 and Site 20 respectively.

During the winter survey (Table 9) the TPC value for Total Dissolved Substances (TDS) was exceeded at two sites, with the highest value, 1530 mg l⁻¹, recorded at Site 14. During the summer survey higher values were recorded with the highest value of 1820 mg l⁻¹ recorded at Site 16. In addition the TPC for TDS was exceeded at four sites during the study period.

TPC values for dissolved oxygen was not set by the KNP but international standards for oxygen saturation is set at between 80 and 120% (Dallas & Day 2004) while Kempster *et al.* (1980) suggested a dissolved oxygen concentration range of 4 to 5,8 mgl⁻¹ as acceptable for aquatic life. During the winter survey (Table 9) oxygen saturation varied considerably with the lowest and highest values measured at sites 13 and 14. In addition, with the exception of sites 4, 9, 19, 20, 21 and 22, the values at all the sites were not within the set parameters. Table 10 shows that during the summer surveys matters improved with oxygen values at eleven of the sites within the parameters and no values above the upper parameter of 120%.

With regard to oxygen concentration the results obtained (Tables 9 and 10) show that in both surveys only four sites were within the set parameters. In general, the oxygen concentrations exceeded the upper limit, which could be an indication of excessive algal growth.

When the phosphorous TPC value (Table 8) is converted to a phosphate value, it can be equated to 0.0326 mg l⁻¹. Table 11 shows that with the exception of the sites where no phosphates could be detected, all the values exceeded the set TPC for phosphates in both the winter and summer surveys. The exceedingly high value of 15mgl⁻¹ obtained at Site 26 is noteworthy.

The highest nitrate value recorded for both the winter and summer was 7 mg l^{-1}, which was recorded at sites 3, 22 and 17 (Table 11). In addition, Site 3 was the only site where the set TPC value was exceeded during both the winter and summer surveys.

The set TPC value for ammonium was exceeded at a number of sites, with values varying between 0 and 46 mg l^{-1} in winter, and 0 to 55 mg l^{-1} in summer (Table 11).

Site no.	Site name	pH	Electrical conductivity	Dissolved oxygen	Dissolved oxygen	Total dissolved substances	Total Suspended Solids	Flow
			μScm^{-1}	%	mg l^{-1}	mg l^{-1}	mg l^{-1}	
1	Tshamidzi							NW
2	Red Ivory							NW
3	Makuleke	7.86	631	76.3	5.73	317	0.07	N
4	Ntlhaveni	8.25	285	97.9	7.8	143	0.1435	L
5	Groot Geluk	6.69	287	72.7	6.35	143	0.18	N
9	Malamulele	7.17	832	96	8.2	417	0.054	N
10	Halahala							NW
11	Halahala Dam	7.74	217	165	12.8	109	0.82	N
12	Mashobye	8.06	759	56.9	4.69	378		N
13	Phugwane East	7.45	269	25.1	2.45	308	0.144	N
14	Dili	7.68	2590	178.4	14.62	1530	0.0795	N
15	Zari	7.73	855	52	4.82	429	0.2	N
16	Mphongolo-Phugwane							NW
17	Giyani bridge	7.94	164.5	68	5.6	82.7	0.045	L
18	Jilongo	9.09	247	149.2	11.24	126	0.11	N
19	Altein	8.76	295	115	9.23	147	0.0795	N
20	Shangoni	8.79	1422	95.3	8.72	710	0.013	L
21	Larini	8.2	1027	93	8.8	517		N
22	Middle Firebreak	8.88	503	117.7	10.15	251	0.087	N
23	Red Rocks							NW
24	Dipene							NW
26	Dzombo	7.9	1243	55.2	4.26	623	0.09	L

Table 9. The *in situ* measured and observed water quality and quantity at the sampling sites in the Shingwedzi River sub-catchment, South Africa, during the winter (May and June) surveys. In the table NW = no water, N = no flow, L = low flow, M = moderate flow and H = high flow (Adapted from Fouché & Vlok, 2010).

Table 12 shows that the calculated Water Quality Index (WQI) scores of the sites where diatoms were collected ranged from "poor" to "good" with none of the sites in either the "excellent" and "very poor" classes. It should be noted that with the exception of Site 14, the sites rated as "good" were in lower percentile of the range of that class and closer to being rated as "medium". The same reasoning could be applied to sites 4, 6 and 20 which lie in the lower percentile of the "medium" class. In totality these results indicate that the water quality within the system is under threat.

At this point the six sites where as many as three TPCs were exceeded (Table 11) should be highlighted. Four of the sites (5, 13, 22 and 25) are within the KNP boundaries while two of the sites (9 and 17) are outside. The calculated WQI scores (Table 12) show that the water quality for sites 5, 9, 25 and 17 is "poor" while the score of the Site 13, which lead to a rating of "medium" lies within the lower percentile of that class and could be regarded as "poor".

Site no.	Site name	pH	Electrical conductivity	Dissolved oxygen	Dissolved oxygen	Total dissolved substances	Total Suspended Solids	Flow
			μScm^{-1}	%	mg l^{-1}	mg l^{-1}	mg l^{-1}	
1	Tshamidzi	8.52	762	119.2	9.35	382	0.053	L
2	Red Ivory	8.02	729	32.6	2.89	365	0.0675	L
3	Makuleke	8.01	1003	53.8	4.7	498	0.1775	L
4	Ntlhaveni	8.43	618	84.4	6.24	390	0.1	L
5	Groot Geluk	8.62	395	100.3	7.84	198	0.194	N
6	Sirheni	7.33	230	47.2	4.7	115	0.09	N
7	Vlakteplaas	8.9	1437	58.2	4.73	718	0.1235	N
8	Shisha/Mphongolo	7.08	271	52.5	3.98	271	0.1155	L
9	Malamulele	6.83	750	60.3	5.04	375	0.122	N
10	Halahala							NW
11	Halahala Dam	6.75	194.3	36.9	3.18	96.6	0.082	N
12	Mashobye	8.15	269	90.8	7.09	133	0.1	N
13	Phugwane east	7.48	326	78.6	6.73	160	0.1175	L
14	Dili	8.06	345	108.6	8.07	178	0.0795	L
15	Zari	9.13	445	100.9	7.3	223	0.183	L
16	Mphongolo/Phug-wane	7.91	366	38.5	3.4	1820		L
17	Giyani bridge	8.0	622	33.6	2.37	312		N
18	Jilongo	8.3	333	72.6	6.7	166	0.23	N
19	Altein	8.2	279	75.8	6.77	140	0.029	N
20	Shangoni	8.91	1029	70.7	6.85	515	0.025	N
21	Larini	7.85	169	86.4	6.32	84.7	0.1275	L
22	Middle Firebreak	8.68	176	91.6	7.02	88.4	0.0665	L
23	Red Rocks	8.53	202	88.0	6.95	120.0	0.124	L
24	Shingwedzi	7.55	339	89.1	6.99	170	0.88	M
25	Dipene	7.91	410	86.8	6.31	207	0.062	L
26	Dzombo	7.59	1278	43.9	3.22	642	0.33	N

Table 10. The *in situ* measured and observed water quality and quantity at the sampling sites in the Shingwedzi River sub-catchment, South Africa, during the summer (February and March) surveys. (In the table NW = no water, N = no flow, L = low flow, M = moderate flow and H = high flow (Adapted from Fouché & Vlok, 2010).

Site no.	Site name	PO_4 mg l^{-1}		NO_2 mg l^{-1}		NO_3 mg l^{-1}		NH_4 mg l^{-1}	
		Summer	Winter	Summer	Winter	Summer	Winter	Summer	Winter
1	Tshamidzi	4		0.6		4		22	
2	Red Ivory	4		0.7		4		20	
3	Makuleke	0		0.7		7	6	0.1	10
4	Ntlhaveni	0.25	0	0.6	0.7	4	3	0	0.1
5	Groot Geluk	0.25	4	0.5	0.6	2	5	20	6
6	Sirheni	3		0.05		2		0	
7	Vlakteplaas	4		0.6		3		0.1	
8	Shisha/ Mphongolo.	4		0.6		2		21	
9	Malamulele	2	4	0.6	0.6	3	3	0.1	25
12	Mashobye	3		0.2		2		22	
13	Phugwane east	3	2	0.7	0.7	2	4	55	21
14	Dili	0	0.25	0.7	0.6	4	2	4	24
15	Zari	0	0	0.7	0.7	2	3	6	22
16	Mphongolo/ Phugwane		0.2		0		1		0
17	Giyani bridge		3		0.6		7		20
18	Jilongo		4		0.6		3	40	0
19	Altein	3	2			2	2	0	0
20	Shangoni	0	0	0	0	0	1	0	38
21	Larini	3		0.6		4		0.3	
22	Middle Firebreak	3	0	0.2	0	7	2	25	23
23	Red Rocks	0.1	3	0.7	0.7	4	4	0.1	0
24	Shingwedzi	3		0.6		3		21	
25	Dipene	3	0.2	0.7	0	2	2	0	46
26	Dzombo	15	0	0.6	0.5	2	2	0.2	44

Table 11. Water quality at the sampling sites in the Shingwedzi River sub-catchment during the February and March (summer) and the May and June (winter) surveys (Adapted from Fouché & Vlok, 2010).

3.5 Biological indices
3.5.1 Diatoms

Table 13 shows the calculated diatom based index scores of the Specific Pollution Sensitivity Index (SPI), the Biological Diatom Index (BDI) and the Trophic Diatom Index (TDI). These results are classified into water quality, ecological status and trophic state classes at the hand of the criteria listed in Table 14.

The allocated water quality classes, based on the SPI and BDI scores, show that with the exception of sites 5 and 14 all the sites were rated as either "unsatisfactory" or "poor" with Site 21 the only site rated as "good". Because of the low SPI and BDI scores the ecological status of all the sites ranged from "poor" to "moderate" quality with no sites rated as "good" quality.

TDI scores are used to determine the trophic status of a water resource and the calculated scores (Table 13) show that a history of high nutrient content, with most sites rated as eutrophic or hypertrophic, was detected in the system. The only sites where low nutrient levels were detected were sites 25 and 21 that were classified as oligotrophic and oligo-/mesosotrophic respectively. The fact that five sites, namely 4, 6, 7, 12 and 24, were rated as hypertrophic should be noted. It is however important to take cognisance of the fact that only Site 4 was outside the KNP and that the trophic status could be ascribed to natural conditions such as a high organic content resulting from animal faeces and organic material of plant origin.

Site numbers	Site name	Water Quality Index scores	Water quality classes (91-100 = excellent, 71 – 90 = good, 51 – 70 = medium, 26 – 50 = poor and 0 – 25 = very poor).
3	Makuleke	76	Good
4	Ntlhaveni	56	Medium
5	Groot Geluk	42	Poor
6	Sirheni	53	Medium
7	Vlakteplaas	60	Medium
9	Malamulele	42	Poor
12	Mashobye	67	Medium
13	Phugwane East	56	Medium
14	Dili	81	Good
17	Giyani bridge	40	Poor
18	Jilongo	76	Good
19	Altein	70	Medium
20	Shangoni	56	Medium
21	Larini	50	Poor
22	Middle Fire Break	74	Good
23	Red Rocks	79	Good
24	Shingwedzi	67	Medium
25	Dipene	48	Poor
26	Dzombo	62	Medium

Table 12. Water Quality Index (WQI) values and the resultant water quality classes (Bate *et al.*, 2004) for sites in the Shingwedzi River sub-catchment, South Africa (Fouche & Vlok, 2009).

3.5.2 Macro-invertebrates

Macro-invertebrates were sampled at 21 sites during summer (Figure 5) and at eleven sites during the winter survey (Figure 8). The results show that not all the required biotopes (Thirion, 2007) were present at all the sites. During the summer survey, ten sites had all the biotopes namely gravel-sand-mud (GSM), stones (S) and marginal vegetation (VEG) present while nine sites had VEG and GSM and two sites only had GSM. Because of the lack of biotopes the calculated SASS results should be viewed with caution.

Site number	Site name	SPI	BDI	TDI
3	Makuleke	8.8	8.3	70.9
4	Ntlhaveni	7.8	8.1	83.8
5	Groot Geluk	12.7	12.9	66.6
6	Sirheni	10.7	10.5	79.7
7	Vlakteplaas	11	9.9	78.5
12	Mashobye	9.3	8.6	83.4
13	Phugwane East	13.1	10.5	68.6
14	Dili	13.4	12.2	72.5
18	Jilongo	11.5	6.3	73.1
19	Altein	13.5	8.8	66.8
20	Shangoni	11	9.8	76
21	Larini	16.3	13.6	45.9
22	Middle Fire Break	13	10.3	67
23	Red Rocks	11	11.2	68.8
24	Shingwedzi	10.3	8.6	86.5
25	Dipene	12.4	13.4	36.5
26	Dzombo	8.6	9.7	74

Table 13. Calculated scores for the Specific Pollution Sensitivity Index (SPI), the Biological Diatom Index (BDI) and the Trophic Diatom Index (TDI) for sites the Shingwedzi River sub-catchment, South Africa (Adapted from Fouché & Vlok, 2009).

SPI and BDI			TDI	
Index scores	Water quality	Ecological status	Index scores	Trophic status
> 17	Very good	High quality	< 35	Oligotrophic
15 - 17	Good	Good quality	35 – 50	Oligo-mesotrophy
12 - 15	Satisfactory	Moderate quality	50 - 60	Mesotrophy
9 - 12	Unsatisfactory	Poor quality	60 - 75	Eutrophic
< 9	Poor		> 75	Hypertrophy

Table 14. Index scores for the calculated diatom indices and the related classification (Adapted from Kelly & Whitton, 1995; Lenoir & Coste, 1996).

During the summer survey, the highest SASS score of 81 was recorded at Mashobye where although only VEG and GSM biotopes were present specimens belonging to fifteen families were identified. Of the fifteen families only four were regarded as sensitive, scoring above 8 on the sensitivity scale. In addition seven families were air breathers, indicating that they could survive a poor water quality environment. Five other sites, namely Red Ivory, Phugwane East, Jilongo, Altein and Shangoni, had a score ranging between 65 and 75, with the number of families recorded varying between twelve and fifteen. These sites all had the full range of biotopes but low numbers of sensitive families and high numbers of air breathing families were observed indicating that the water quality was poor, probably due to organic pollution. When the SASS results for the sites where all the biotopes were present for the summer survey (Figure 5) are considered a decreasing downstream trend in water quality in the Mphongolo, Phugwane and Shingwedzi rivers, is observed. Similar results were observed during the winter survey (Figure 6) when the biotope diversity was lower with most of the sites having only stones (S) as a biotope. The lowest score of 14 was recorded at Zari where GSM was the only available biotope compared to the highest score of 72 recorded at Larini, again with stones (S) as the only available biotope. No sensitive families were present at Zari, whilst two of the four families present were air breathers. At Larini four sensitive families were recorded, with five air breathing families present.

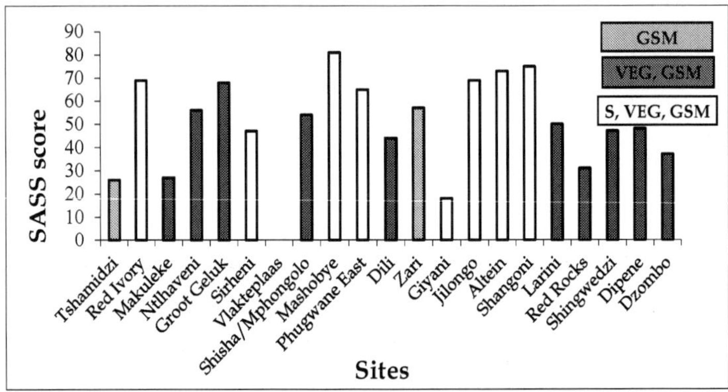

Fig. 5. SASS scores for the summer survey in the Shingwedzi River sub-catchment, South Africa. (Biotopes: GSM = ground, sand and mud; VEG = vegetation; S= sand).

According to Dallas & Day (2004) the two factors that have the most profound impact on macro-invertebrate populations are flow and the amount of total dissolved substances. Low SASS scores are generally regarded as an indication of poor quality (Thirion, 2007). Based on the scores obtained, and supported by the low numbers of sensitive families and the high number of air breathers, it is clear that the water quality in the Shingwedzi River system is poor. It can be ascribed to organic pollution which result from the lack of infrastructure in the catchment, with no or dysfunctional sewage treatment works and cattle faeces the main contributors. The Shingwedzi River System can be classified as an intermittent seasonal river (Rossouw *et al.*, 2005). These authors describe this type of river as "a river that exhibit seasonally predictable intermittent flow where surface flow may disappear for a period each year reducing the channels to isolated pools or drying up completely during the dry season and flow that commence during the rainy season and may be sustained or be intermittent over the wet season". In addition these intermittent rivers can have variable flow and one to two year dry cycles in a five-year period are normally present (Rossouw *et al.*, 2005). The low SASS scores could therefore be a result of a lack of flow, which is exacerbated by the weirs that obstruct flow, and an increase in water extraction.

3.5.3 Vegetation

Table 15 shows that with regard to the calculated Vegetation Response Assessment Index (VEGRAI) scores and the resultant Ecological Categories the sites ranged fom a low category D, which reflects a "largely modified" ecosystem (Table 16) to a category A, which reflects an "unmodified or natural" system (Table 16). Although there is clear indication that impacts are negatively affecting the riparian zone more than 50% of the sites were in a better state than a category C. This indicates a system that is moderately modified to largely natural.

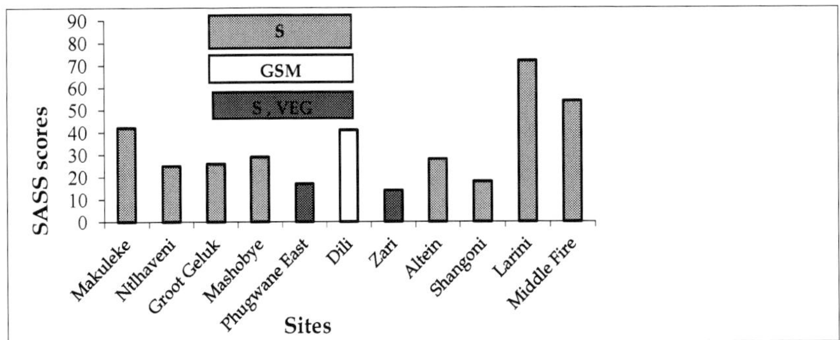

Fig. 6. SASS scores for the winter survey in the Shingwedzi River sub-catchment, South Africa. (Biotopes: GSM = ground, sand and mud; VEG = vegetation; S= sand).

Site no	Site Name	River	VEGRAI score	EC class
1	Tshamidzi	Mphongolo	72.4	C
2	Red Ivory	Mphongolo	73.3	C
3	Makuleke	Mphongolo	68.9	C
4	Ntlhaveni	Mphongolo	80	C/B
5	Groot Geluk	Mphongolo	81.5	B
6	Sirheni	Mphongolo	76.7	C/B
7	Vlakteplaas	Shisha	94.1	A
8	Shisha-Mphongolo confluence	Mphongolo	78.6	C/B
9	Malamulele	Phugwane	49.0	D
10	Halahala	Phugwane	52.4	D
12	Mashobye	Phugwane	79.4	C/B
13	Phugwane-east	Phugwane	82.8	B
14	Dili	Phugwane	76.8	C
15	Zari	Phugwane	78.5	C/B
17	Giyani bridge	Shingwedzi	63.4	C
18	Jilongo	Shingwedzi	77.4	C/B
19	Altein	Shingwedzi	61.7	C/D
20	Shangoni	Shingwedzi	80.1	B/C
21	Larini	Shingwedzi	72.4	C
22	Middle Firebreak	Shingwedzi	60.5	C/D
23	Red Rocks	Shingwedzi	81.2	B/C
24	Shingwedzi camp	Shingwedzi	75.4	C
25	Dipene causeway	Shingwedzi	80.6	B/C
26	Dzombo	Dzombo	90.7	A/B

Table 15. The calculated Vegetaton Response Assessment Index (VEGRAI) scores and Ecological Category (EC) classes of the sites surveyed in the Shingwedzi River and tributaries, South Africa (Adapted from Fouché & Vlok, 2009).

Index scores (% of total)	Category/class description	Ecological category / Ecological class
90 - 100	Unmodified, natural.	A
80 - 89	Largely natural with few modifications. A small change in natural habitats and biota may have taken place but the ecosystem functions are essentially unchanged.	B
60 - 79	Moderately modified. Loss and change of natural habitat and biota have occurred, but the basic ecosystem functions are still predominantly unchanged	C
40 - 59	Largely modified. A large loss of natural habitat, biota and basic ecosystem functions has occurred.	D
20 - 39	Seriously modified. The loss of natural habitat, biota and basic ecosystem functions is extensive.	E
0 - 19	Critically/Extremely modified. Modifications have reached a critical level and the system has been modified completely with an almost complete loss of natural habitat and biota. In the worst instances the basic ecosystem functions have been destroyed and the changes are irreversible.	F

Table 16. The generic Ecological Categories, or classes, and their descriptions related to VEGRAI and FRAI scores (Adapted from Kleynhans & Louw, 2007).

The vegetation at seven sites in the Mphongolo River was surveyed in this study. The VEGRAI results (Figure 7) show that the three sites outside the KNP boundaries, Tshamidzi, Red Ivory and Makuleke, were in the worst condition (Ecological Category C). At Nthlaveni limited improvement was observed with the Ecological Category improving towards a B category (EC = C/B). This was followed by a further improvement to a B category at Groot Geluk. The presence of the Sirheni Dam however had a negative influence on the vegetation at the Sirheni site, which is downstream of the impoundment, resulting to a change to a category C. Some improvement was again observed downstream of this site, which could be partly ascribed to the influence of the Shisha River and the vegetation at the Mphongolo-Shisha confluence site was classified as a category C/B. The fitted trend line in figure 9 indicates a downstream improvement in the state of the riparian vegetation along the Mphongolo River.

The Tshamidzi site is in the upper catchment of the Mphongolo River and lies on the watershed between the Luvuvhu and Shingwedzi catchments. Eighteen woody species were recorded of which only ten were classified as "riparian". There is therefore a degree of terrestrial invasion in the form of the *Dichrostachys cinerea* recorded and this "invasion" is typical of an area with low flow and variable seasonal differences. The population structure of the riparian woody plants seems to be in good condition as there are sufficient numbers of juveniles, e.g. *Breonadia salicina*, present. This species is regarded as a TPC species indicative of healthy riparian vegetation. The Red Ivory site is characterized by a deep V-shaped valley in which the river flows and this has the effect that the riparian zones on both sides are narrow. One of the implications is that very few shrubs can establish and this could also be the reason for the low diversity with only nine species of woody plants recorded. The actual site is an unused river crossing with resultant erosion. This is exacerbated by the fact that the surrounding area is used for cultivated lands and grazing. The Makuleke site is influenced by the imoundment immediate upstream of the survey site. As a result of the effect of flood control by the dam wall a number of woody terrestrials have established and these species contribute up to 30% of the recorded plants at the site. On the other hand seepage, resulting from return flow, has led to the establishment of a well defined marginal zone. As is the case with rivers that flow through inhabited areas, signs of pollution are evident in the form of excessive algal growth and the presence of solid waste such as plastic bags and other household refuse.

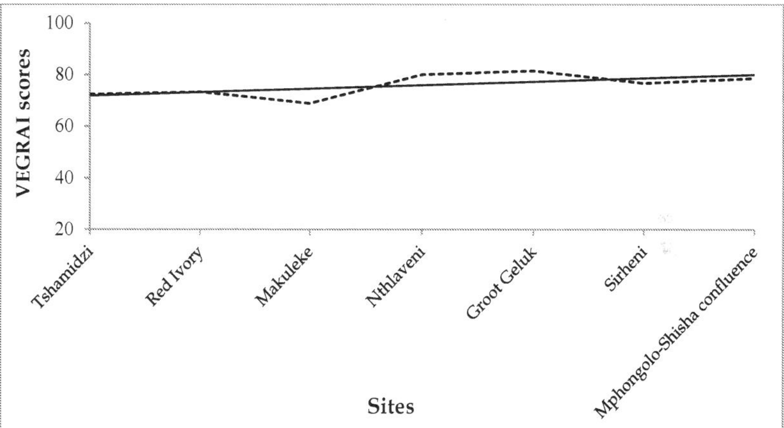

Fig. 7. Calculated VEGRAI scores for sites 1, 2, 3, 4, 5, 6 and 8 in the Mphongolo River, South Africa. (The solid line represents a fitted trend line). (Adapted from Fouché and Vlok, 2009)

The Shisha River is unique because it originates within the KNP and therefore does not have as many of the anthropogenic impacts as the other rivers in the system. With the exception of the Langtoon Dam in its upper catchment the impacts observed are all natural. The Vlakteplaas site, which is typical of this reach of the Shisha River, consists of a deep stagnant pool that gives permanency to the site resulting in a well established marginal zone or wetted perimeter. It is possibly the most undisturbed site of all the sites surveyed during this study and was classified as the only category "A" site of the survey.

In the Phugwane River the vegetation was surveyed at six sites and the VEGRAI results (Figure 8) indicate that two of the sites, namely Malamulele and Halahala which are both outside the KNP boundaries, are the only two sites in the entire system that were classified as a category D, which reflects that they are "largely modified" (Table 16). In the case of the Malamulele site it can be attributed to human interference in the form of farming activities and pollution. Downstream of this site some improvement was observed at the Mashobye site. As was the case with the Mphongolo River the downstream trend is an upwards one in the riparian vegetation scores. The observed trend throughout the river is depicted by the trend line fitted in Figure 8.

The Malamulele site is definitely the most impacted site of all the sites surveyed in the study. It is situated at origin of the stream flowing out of the wetland. This site in the headwaters of the river is in an area that is poorly managed with crop fields extending into the riparian zone. The Halahala site is a typical example of the effect of no or very little water flow and water quantity and seasonality can be regarded as the factor with the most influence on the classification of the vegetation. Although thirteen woody species were recorded in the riparian zone only five were classified as riparian. This indicates an invasion of terrestrial species into the riparian zone which is typical of ephemeral streams and in particular those that flow through low rainfall areas. In general the population structure of the woody riparian species seems in order but very few juveniles were observed. In addition no non-woody riparian species were observed in the marginal zone.

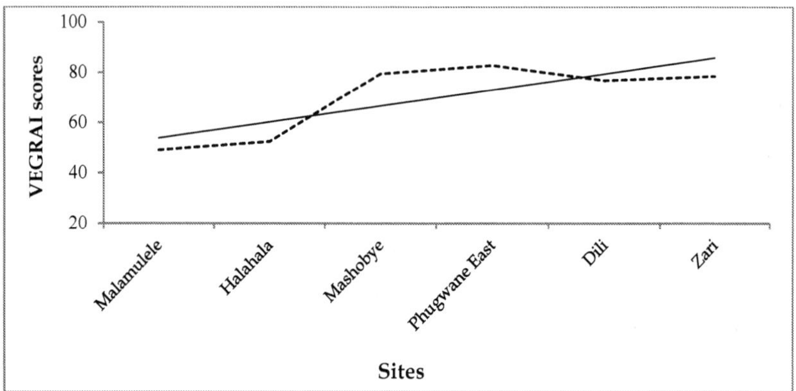

Fig. 8. Calculated VEGRAI scores for sites 9, 10, 12, 13, 14 and 15 in the Phugwane River, South Africa (The solid line represents a fitted trend line). (Adapted from Fouché & Vlok, 2009).

The vegetation at nine sites in the Shingwedzi River was surveyed during this study. As was the case with both other rivers a downstream increase in the status of the riparian vegetation was observed as depicted by the fitted trend line in Figure 9. The severe fluctuation between sites should be noted. In this river no unmodified or natural sites were observed.

The Altein site shows distinctive signs of a high degree of impact in the form of wood cutting and grazing by cattle and goats. As a result the grass cover consisted of pioneer species. The broken road bridge acts as a hydraulic control forming a deep pool immediately upstream. Further upstream of the site an extraction weir regulates the flow in the river. Although 16 woody species were recorded five species, or more than 30%, are terrestrial species. In addition the woody species is dominated by shrubs (e.g. *Gymnosporia senegalensis*) and creepers (e.g. *Acacia ataxacantha*). Very few large tree specimens were present which can be the result of wood harvesting. It is however important that some juveniles of *Philenoptera violacea* and sub-adult *Ficus sycomorus* were observed. Both these species are regarded as preferential riparian woody species

The vegetation at the Shangoni site, which is downstream from the abandoned mines with resulting bad water quality, is still reasonably intact. In part this can be attributed to the remoteness of the site and the resultant low human impact. Fields that were historically ploughed and grazing areas were observed near the site but the effects are not evident at the site. In addition there are also few humans living upstream of the site. On the other hand the low impact on the vegetation can be ascribed to the fact that the adverse effect of the water quality resulting from drainage of the mines are "more recent" and the riparian zones, other than the marginal zone have not yet been affected. The adverse effects in the marginal zone are evident from the fact that despite of the semi-permanent pool only a few scattered sedges are present at the site.

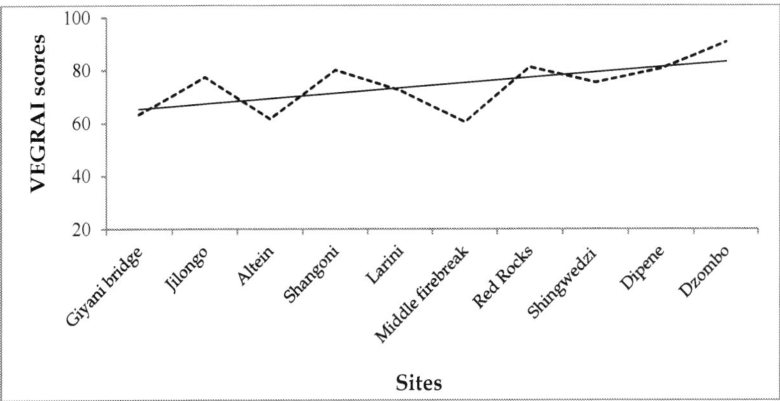

Fig. 9. Calculated VEGRAI scores for the sites in the Shingwedzi River, South Africa (The solid line represents a fitted trend line). (Adapted from Fouché & Vlok, 2009).

It is important to note that the Shingwedzi site, downstream of the Rest camp, is impacted by subsurface water extraction, polllution and a low-water bridge. This bridge forms a weir and a large semi-permanent pool is present at the site. The river has a wide sand bed, in excess of 80m, with the pool and deeper channel against the left hand bank. The low diversity of woody riparian species is of concern but on the other hand no invasion by terrestrial or exotic species was recorded. A further concern is the low juvenile count of the riparian species present.

The Dzombo site is the second least impacted site and is in the least impacted river in the current survey. It is however important to note that it is a small ephemeral stream which is subjected to extended periods of drought. This is reflected by the high number terrestrial species present in the riparian zone.

3.6 Fish
3.6.1 Historic fish distribution and biodiversity of the Shingwedzi River

Fish distribution data were primarily obtained from the data bases of the Limpopo Department of Environment and Tourism (LEDET), the South African Institute of Aquatic Biodiversity (SAIAB), the KNP, the now defunct Transvaal Provincial Administration (TPA) as well as from the work of Gaigher (1969), Pienaar (1978), Russell (1997) and Olivier (2003). In addition data was obtained from experts (Angliss, pers com.[1]; Deacon, pers com.[2]; Kleynhans, pers com.[3]; Engelbrecht, pers com. [4]) and added on to the data. This data were then used to construct the historic data and eventually deduct the reference state (RS).

---

[1] Angliss, M.K., Anglo Platinum, Polokwane, South Africa.

[2] Deacon, A.R., Scientific Services, Kruger National Park, South Africa

[3] Kleynhans, C.J. Resource Quality Services, Department of Water Affairs, Pretoria, South Africa.

When the historic data was compared with the fish data provided by Kleynhans *et al.* (2007a) for the two reference fish distribution sites (2LF2 and 2LF3) in the Shingwedzi River system some differences are observed namely:

a. That although Kleynhans *et al.* (2007a) lists the eel *Anguilla marmorata* no reference to its presence was found in the other literature surveyed.

b. That the Southern Mouth Brooder, *Pseudocrenilabrus philander,* is listed by Kleynhans *et al.* (2007a) as "a species derived to be present" at site 2LF3 in the Mphongolo River, the only record in the surveyed literature is that of Pienaar (1978) who collected the species in the vicinity of Red Rocks..

c. That the tigerfish, *Hydrocynus vittatus* and the Purple Labeo, *Labeo congoro*, are not listed at both 2LF2 and 2LF3. This despite the fact that the presence of *H. vittatus* is reported by Pienaar (1978) at sites in the proximity of the current sites 6, 15, 24 and 25 and by Deacon (pers com.[2]) at Site 6.

Family	Historic biodiversity	Recorded during the survey
Characidae	*Hydrocynus vittatus* *Brycinus imberi* *Micralestes acutidens*	*Brycinus imberi*
Cyprinidae	*Labeobarbus marequensis* *Barbus afrohamiltoni* *Barbus matozzi* *B. trimaculatus* *B. paludinosus* *B. unitaeniatus* *B. viviparus* *B. annectens* *B. toppini* *B. radiatus* *Labeo congoro* *L. cylindricus* *L. rosae* *L. ruddi* *L. molybdinus* *Opsaridium peringueyi* *Mesobola brevianalis*	*Barbus afrohamiltoni* *B. trimaculatus* *B. unitaeniatus* *B. viviparus* *B. annectens* *B. toppini* *B. radiatus* *L. cylindricus* *L. rosae* *L. ruddi* *L. molybdinus* *Mesobola brevianalis*
Schilbeidae	*Schilbe intermedius*	*Schilbe intermedius*
Clariidae	*Clarias gariepinus*	*Clarias gariepinus*
Mochokidae	*Synodontis zambezensis* *Chiloglanis paratus*	
Cichlidae	*Oreochromis mossambicus* *Tilapia rendalli* *Pseudocrenilabrus philander*	*Oreochromis mossambicus* *Tilapia rendalli*
Gobiidae	*Glossogobius giuris*	*Glossogobius giuris*
Anguillidae	*Anguilla mossambica*	

Table 17. The historic fish diversity and fish recorded during this survey in the Shingwedzi River sub-catchment, South Africa.

---

[4] Engelbrecht, J. Private Limnology Consultant, Lydenburg, South Africa.

New frequency of occurrence (FROC) data was constructed by combining the findings of the Kleynhans *et al.* (2007) with the historic data gathered in this project. As part of this exercise certain sites were grouped together with the existing reference sites or were grouped on their own.

3.6.2 Results of the survey

Table 17 shows that during the surveys a total of 18 fish species, belonging to six families, were collected throughout the river system. Although specimens from of eleven species were not collected the absence of the *Labeobarbus marequensis, Hydrocynus vittatus* and the two eel species (*Anguilla marmorata* and *A. mossambica*) is of concern since both *H. vittatus* and *L. marequensis* is dependent on flow for breeding and the eels are the only true migrators that are negatively affected by barriers that prevent migration. The absence of *Barbus matozzi* agrees with the observations of Engelbrecht (pers com.[4]) who found the species absent in areas of historic distribution in other rivers in adjoining WMAs. The absence of Purple Labeo (*Labeo congoro*) and the Brown Squeaker (*Synodontis zambezensis*) specimens in the samples could be ascribed to the lack of deep pools. The fact that no *Barbus paludinosus* specimens were collected is ascribed to a lack of habitat, which includes shallow water adjacent to the wetted perimeter. When flow is diminished this is the type of habitat primarily affected. The absence of *Micralestes acutidens* is of real concern as this is one of the more common species, as is the case with *L. marequensis* . The specimens of the River Sardine (*Mesobola brevianalis*) and the Redeye Labeo (*Labeo cylindricus*), collected at sites 3 and 17 respectively, are the first records of the two species at the specific sites.

Two approaches were followed in the calculation of the FRAI scores. In the first approach each site was treated as an individual data point in order to establish the extent of anthropogenic impacts between sites. The second approach was to combine the survey results of the sites within the same ecoregion of each tributary. This resulted in groupings (Table 5) which each was then regarded as a reach within the river system (Kleynhans, 2007). For each of these reaches a FRAI score was then calculated.

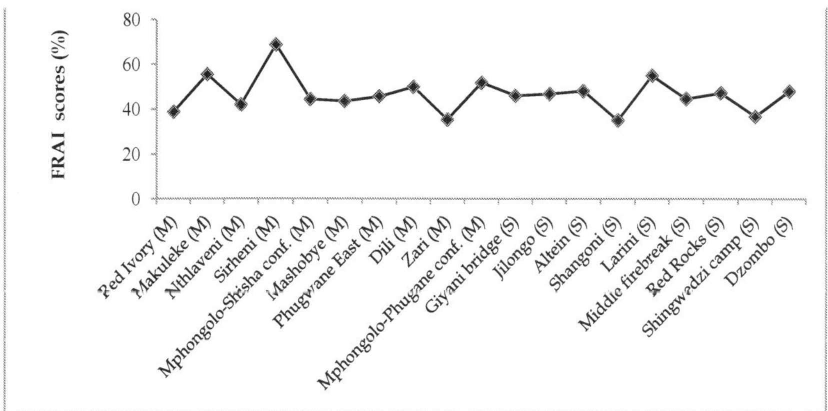

Fig. 10. The calculated Fish Response Assessment Index (FRAI) scores for the sites in the Shingwedzi River sub-catchment, South Africa (Adapted from Fouché & Vlok, 2010).

The FRAI results obtained for individual sites in the main stem of the river are shown in Figure 10. In the tributaries a trend of downstream improvement of the FRAI scores were observed. The exception was the low score obtained at Zari, in the Phugwane River, which indicates that a negative impact had occurred upstream of the site. Initially an upward trend was observed in the main stem Shingwedzi River, but a distinct decrease in the FRAI score occurred at Shangoni, below which a downward trend was observed.

Table 18 shows the FRAI scores for four of the reaches formed by the grouping of sites in the sub-level 2 ecoregions. Because Site 1 (Tshamidzi) had received water only one day prior to the March 2008 survey and Site 2 (Red Ivory) only had a stagnant pool Reach I was not included. The FRAI scores show that the reaches were in a reasonably good ecological state, with the Mphongolo River being the best. The Phugwane River, with a score of 60,8 was in the worst state and this could be ascribed to the fact that this tributary had the lowest and most erratic flow. Notably, the sites within reaches II, III and IV comprised a combination of sites lying outside and inside the boundaries of the KNP.

Reach	River/Tributary	Calculated FRAI score	Ecological category (EC)
II	Mphongolo and Shisha	78.7	B/C
III	Phugwane	60.8	C/D
IV	Shingwedzi	70.5	C
V	Shingwedzi	62.2	C

Table 18. The FRAI scores calculated for the ecological reaches in the Shingwedzi River and tributaries, South Africa. (Fouché & Vlok, 2010).

4. Conclusion

A review of the literature, as cited in this report, has shown that the perception exists that phosphate is the only water quality parameter adversely affected by activities within the sub-catchment and creates the idea that this river has low levels of pollution. This study has however shown that it was not only phosphate but that it also applied to the nitrogenous compounds, ammonium and nitrate. All three the compounds exceeded the TPC values set by the KNP at a number of sites. In addition the TPC values for pH, electrical conductivity and Total Dissolved Substances were also exceeded in a number of instances. Based on the above the calculated Water Quality Index scores shows that on average the water quality of the rivers could only be rated as of "medium" quality. These findings are supported by the low SASS and diatom index scores and are proof that the level of pollution is higher than originally thought. The fact that this is underpinned by the diatom based indices indicates that the pollution is a continuous and ongoing process.

It is of concern that the high nutrient content in the water and the resultant trophic status of the rivers, as is indicated by the high Trophic Diatom Index scores, is most certainly the result of anthropogenic impacts such as incorrect or lack of waste management practices and return flow from commercial agricultural. With an increase in the population, as is expected in the sub-catchment, these activities, and consequently pollution, will increase.

Research for this project has shown that historically the mean annual run-off (MAR) in the sub-catchment is low and consequently the flow in the Shingwedzi River and tributaries is not only low and episodic but also extremely variable. The current study however points to even lower flows. The evidence for this is observed in the riparian vegetation where there is a visible encroachment by terrestrial woody species. Lower water levels during floods as well as a decrease in the base flow create a suitable environment that allows these woody terrestrial species to flourish and even become dominant in some instances. The second indication that flow has diminished lies in the absence of fish species that are dependent on flow for breeding such as the tigerfish, *H. vittatus,* and the Lowveld largescale yellowfish, *L. marequensis* (Fouché, 2009). In addition the fragmentation of the rivers caused by the building of weirs has destroyed the connectivity and consequently the loss of the only true migratory species within the system namely the eel *A. mossambica.*

Higher demands for water resulting from an ever increasing population will ultimately lead to the construction of more weirs to create water storage reservoirs. This will not only diminish the flow and lead to further fragmentation of the river but incorrect management of releases could negatively impact on the seasonality and flow regime of the river. This in turn will not only affect the reproductive strategies of instream biota such as fish but also the recruitment of the obligate riparian plant species. These weirs can act as migratory barriers and could prevent local migratory fish species to migrate for breeding and dispersion purposes.

Based on the above it can be concluded that the Shingwedzi River and its tributaries is not only vulnerable but increasingly under threat. The vulnerability is related to the fact that these rivers are in a water scarce area. To conserve these rivers, and in particular the role that they play within the landscape, is of utmost importance and can only be achieved through a well designed management plan that is properly implemented and maintained.

5. References

Archibald, R.E.M. (1972). Diversity in some South African diatom associations and its relation to water quality. *Water Res.* Vol. 6, pp1229-1238.

Bate, G.C.; Smailes, P.A. & Adams, J.B. (2004). A water quality index for use with diatoms in the assessment of rivers. *Water SA,* Vol. 30, pp493-498.

Birks, H.B.J.; Line, J.M.; Juggins, S.; Stevenson, A.C. & ter Braak, J.F. (1990). Diatoms and pH reconstructions. *Philosophical Transactions of the Royal Society of London,* B, pp327: 263-278.

Breen, P. (1998). Diatoms, ecosystem health assessment and water quality biomonitoring. In Newell, P. (Ed.), *Proceedings of the First Australian Diatom Workshop,* Deankin University, Australia.

Dallas H.F. & Day J.A. (2004). The effect of water quality variables on aquatic ecosystems: a review. *WRC Report No TT 224/04,* Water Research Commission, Pretoria, South Africa.

Department of Water Affairs and Forestry (DWAF). (1986). *Management of Water Resources of the Republic of South Africa.* Government Printer, Pretoria, South Africa

Department of Water Affairs and Forestry (DWAF). (2004a). *National Water Resource Strategy* (1st Ed). Publication of the Department of Water Affairs and Forestry, Pretoria, South Africa.

Department of Water Affairs and Forestry (DWAF). (2004b). Luvuvhu/Letaba Water Management Area: Internal Strategic Perspective. Prepared by Goba Moahloli Keeve Steyn (Pty) Ltd in association with Tlou and Matji, Golder Asssociates Africa and BKS on behalf of the Directorate: National Water Resources Plannng. *DWAF Report No. P WMA 02/000/00/0304.*

Dickens C.W.S & Graham P.M. (2002).The South African Scoring System (SASS) Version 5 Rapid Bioassessment Method for Rivers. *Afr. J. of Aquatic Science,* Vol. 27, pp1 -10.

Fouché, P.S.O. (2009) Aspects of the ecology and biology of the Lowveld largescale yellowfish (*Labeobarbus marequensis*) (Smith 1843), in the Luvuvhu River, Limpopo River System, South Africa. Unpublished PhD. Thesis, University of Limpopo, South Africa.

Fouché, P.S.O. & Vlok, W. (2009). A baseline survey of the Shingwedzi River Catchment – alien plant infestation, aquatic, geomorphology, and riparian zone integrity. *Final report to the Research and Publications Division, University of Venda, South Africa.*

Fouché, P.S.O. & Vlok, W. (2010). Water quality impacts on the instream biota of the Shingwedzi River, South Africa. *Afr. J. of Aquatic Science,* Vol. 35, no 1, pp 1 – 11.

Gaigher, I.G. (1969). *Aspekte met betrekking tot die Ekologie, Geografie en Taksonomie van Varswatervisse in die Limpopo- en Incomatiriviersisteem.* Unpublished Ph.D, Randse Afrikaanse Universiteit, Johannesburg, South Africa.

Grant, R. & Thomas, V. (2000). *SAPPI tree spotting: Bushveld.* Jacana Education (Pty) (Ltd), Johannesburg.

Grant, R. & Thomas, V. (2001). *SAPPI tree spotting: Lowveld* (2nd ed.) Jacana Education (Pty) (Ltd), Johannesburg.

Kelly, M.G & Whitton, B.A. (1995). The Trophic Diatom Index for Monitoring Eutrophication in rivers. *J. Appl. Phycology,* Vol. 7, pp 1573-1576.

Kemper, N.P. (2001). Riparian Vegetation Index. *WRC report 850/3/01.* Water Research Commission, Pretoria.

Kempster, P.L.; Hatting, W.A.J. & Van Vliet, H.R. (1980). *Summarised water quality criteria.* Report of the Department of Water Affairs, Forestry and Conservation, Pretoria.

Kleynhans, C.J. (2007). Module D: Fish Response Assessment Index in River EcoClassification: Manual for EcoStatus Determination (version 2. Joint Water Research Commission and Department of Water Affairs and Forestry report. *WRC Report No. TT330/08.* Water Research Commission, Pretoria, South Africa.

Kleynhans, C.J. & Louw M.D. (2007). Module A: EcoClassification and EcoStatus determination in River EcoClassification: Manual for EcoStatus Determination (version 2). Joint Water Research Commission and Department of Water Affairs and Forestry report. *WRC Report No. TT 329/08.*

Kleynhans, C.J., Louw, M.D. & Moolman, J. (2007a). Reference frequency of occurrence of fish species in South Africa. Report produced for the Department of Water Affairs and Forestry (Resource Quality Services) and the Water Research Commission. *WRC Report No TT 331/08.*

Kleynhans, C.J.; MacKenzie, J. & Louw, M.D. (2007b). Module F: Riparian Vegetation Response Assessment Index in River EcoClassification: Manual for EcoStatus Determination (version 2). Joint Water Research Commission and Department of Water Affairs and Forestry report. *WRC Report No. TT 333/08.*

Kleynhans, C.J. & Hill, L. (1999). *Ecoregional typing.* Report prepared for the Institute for Water Quality Studies, Department of Water Affairs and Forestry, Pretoria, South Africa.

Kruger National Park (KNP) (2009) Kruger National Park biodiversity management programme: Strategic Adaptive Management. SANParks, Skukuza, South Africa.

Lecointe, C.; Coste, M. & Prygiel, J. (1993). Omnidia: software for taxonomy, calculation of diatom indices and inventories management. *Hydrobiologia,* Vol. 269/270, pp 509-513.

Lenoir, A. & Coste, M. (1996). Development of a practical diatom index of overall water quality applicable to the French National Water Board network. In: Whitton, B.A and Rott, E (Eds). *Use of Algae for Monitoring Rivers II.* Institut. Fur Botanik. Universitat Innsbruck. .

Midgley, D.C.; Pitman, W.V. & Middleton, B.J. (1994). Surface water resources of South Africa 1990. *WRC Report no 298/1.1/94.* Water Resource Commission, Pretoria.

Mucina, L. & Rutherford, M.C. (Eds.) (2006). The vegetation of South Africa, Lesotho and Swaziland. *Strelitzia* 19. South African Biodiversity Institute, Pretoria.

Olivier, L. (2003). The effectiveness and efficiency of the Kanniedood Dam fishway in the Shingwedzi River, Kruger National Park, South Africa. Unpublished MTech. dissertation, Technikon Pretoria.

Passy, S.I.; Kociolek, J.P. & Lowe, R.P. (1997). Five new *Gomphonema* species (Bacillariophyceae) from rivers in South Africa and Swaziland. *J.Phycol.,* Vol. 33, pp 455 – 474.

Pienaar, U. de V. (1978). *The freshwater fishes of the Kruger National Park.* Sigma Press, Pretoria.

Rossouw, L.; Avenant, M.F.; Seaman, M.T.; King, J.M.; Barker, C.H.; Du Preez, P.J.; Pelser, A.J.; Roos, J.C.; Van Staden, J.J.; Van Tonder, G.J. & Watson, M (2005). Environmental water requirements in non-perennial systems. *WRC report 1414/1/05.* Water Research Commission, Pretoria.

Rowntree, K. M. & Wadeson, R. (1999). An index of stream geomorphology for the assessment of river health. Unpublished field manual for channel classification and condition assessment.

Rowntree, K.M.; Wadeson, R.A. & O'Keeffe, J. (2000). The development of a geomorphological classification system for the longitudinal zonation of South African rivers. *South African Geographical Journal,* Vol. 82 no. 3, pp163-172.

Russel, I.A. (1997). Monitoring the conservation status and diversity of fish assemblages in the major rivers of the Kruger National Park. Unpublished Ph.D. Thesis, University of the Witwatersrand, Johannesburg.

Schmidt, E.; Lötter, M. & McCleland, W. (2007). *Trees and shrubs of Mpumalanga and Kruger National Park.* Jacana Media, Johannesburg.

Schoeman, F.R. (1982). The diatoms of the Jukskei-Crocodile River system (Transvaal, Republic of South Africa): A preliminary checklist. *Jl S. Afr.Bot.,* Vol. 48, no. 3, pp 295 – 310.

Strydom, W.F.; Hill, L. & Eloff, E. (2006). Achievements of the River Health Programme 1994 – 2004: A national perspective on the ecological health of selected South African rivers. *Report of the National Aquatic Ecosystem Health Monitoring Programme (River Health Programme).*

Skelton, P.H. (2001). *A Complete Guide to the Freshwater Fishes of Southern Africa.* (2nd Edition). Southern Book Publishers, Halfway House.

Taylor, J.C.; Janse van Vuuren, M.S. & Pieterse, A.J.H. (2007a). The application and testing of diatom-based indices in the Vaal and Wilge Rivers, South Africa. *Water SA.*, Vol. 33, no. 1, pp51-59.

Taylor, J.C.; Harding, W.R. & Archibald, C.G.M. (2007b). An Illustrated Guide to Some Common Diatom species from South Africa. *WRC report TT 287/07*. Water Research Commission, Pretoria

Thirion, C. (2007). Module E: Macroinvertebrate Response Assessment Index in River EcoClassification: Manual for EcoStatus Determination (version 2). Joint Water Research Commission and Department of Water Affairs and Forestry report. *WRC Report No. TT 332/08.*

Van Wyk, B & Van Wyk, P. (1997). *Field guide to trees of Southern Africa.* Struik Publishers, Cape Town.

Watanabe, T.; Asai, K. & Houki. K. (1988). Numerical Water quality monitoring of organic pollution using diatom assemblages. In: Round, F.E. (Ed) *Proceedings of the ninth International Diatom Symposium, Bristol*, August 24-30. Biopress, Bristol & Koeltz, Koenigstein.

8

Natural Materials for Sustainable Water Pollution Management

Kenneth Yongabi, David Lewis and Paul Harris
School of Chemical Engineering, The University of Adelaide
South Australia

"Once you eliminate the Impossible, whatever remains, no matter how improbable, must be the truth" (Sherlock Holmes. (Sir Arthur Conan, Doyle, 1859 - 1930))

1. Introduction

The world Health Organization has estimated that up to 80% of all diseases and sicknesses in the world are caused by inadequate sanitation, polluted water or unavailability of water. (Cheesbrough, 1984, Pritchard et al., 2009 and Yongabi et al., 2010) Faeces, gabbage resulting from improper sewage disposal are an important source of pathogenic organisms in water, especially the causative agents of diarrhoeal and dysentery diseases. Faeces are attractive to flies which support the development of the larval stages (maggots) of filth flies. These hazards couple with the indiscriminate disposal of faeces can also constitute a grave nuisance from the offensive sight and smell. In many parts of rural Africa, toilets and garbage disposal pits and/or sites are cited close to wells.The leachates from these could contaminate ground and surface water (Yongabi et al., 2011) Diseases associated with water could be broadly categorised into five epidemiological groups viz: Waterborne infections e.g cholera, typhoid, infective hepatitis, Water shortage diseases e.g. skin infections, trachoma, Water-impounding diseases e.g schistosomiasis and guinea worm, Water-arthropod disease e.g. malaria onchocerciasis, Chemical constituents either excess or shortage e.g. fluoride - this may have an indirect effect in the body.The World population is currently growing at an unprecendented rate, and as of 1996 a 1.8% growth rate per annum which translates to over 80 million people a year was reported (UN, 1996). There is no doubt that most African countries are presently characterised by inexorable population explosion (Adegbola, 1987 and Yongabi et al., 2010) This has dire consequences to the food and environmental resources, more industries are springing up and the quest for survival is generating a lot of pollution.Inorder to meet the needs of this soaring population, the production capacity in all the sectors would have to be multiplied. All these sectors would need water as a raw material and apparently, there is water shortage in Sub Saharan Africa. For instance, in Nigeria, alot of volumes of water is used and needed for irrigation especially in the dry season.
The potential irrigation area of Nigeria stands at about 2.5 Million hactares which is capable of producing close to 40% of the current total annual crop production. The small scale fadama irrigation constitutes 90% of the country's irrigation potentials (Dada et al., 1990).

This picture is similar across sub Saharan Africa. Apart from agriculture, the health sector, hospitals and pharmaceutical companies in Nigeria uses a lot of water during manufacturing and cooling systems. The waste water generated is high and often discharged untreated. This ultimately pollutes surface water. Such waste water is difficult and`expensive to treat and re-use as it contains a mix of chemical compounds. The major industries in Nigeria like food-based industries and breweries and refineries really use huge volumes of water and as such generate so much waste water.

Taking the brewing industry as a good first point of call, only 8% of the nutrients in the spent grain are used. The other 92% is waste and usually discharged into the environment.Imagine 18.5 hl of wastewater (alkaline is leashed out to the environment to produce a bottle of beer (Ihl) with 1.2kg of BOD, equally, the water input is high. (20hl) Many breweries across sub saharan africa and the world over function along similar lines (Pauli, 1998) These sectors depend heavily on water which is scare and generates a lot of wastewater that is at best untreated and goes along way to generate ecological imbalances especially in the face of population explosion. As consequence of population explosion is heavy dependence on fresh water resources, fresh water gradually becoming impoverished in many parts of the world through a number of means; contamination, reclaimation and exhaustion. For instance, in Jordan and Yemen 30% of their water from their ground water acquifers are depleted per annum than the acquifers are able to recharge (Engle/Manand /eroy, 1993). This trend may be similar across many countries the world over.

1.1 Statement of problem / justification

Estimates reveal that water borne diseases contribute to the death of 4 million children in the developing countries each year. This estimate by UNICEF may be a far under estimation of the real situation on ground. In many African countries for instance, water is scarce in both the rural and urban settings. A local survey carried out in the states of Bauchi, Plateau and Benue states shows that most of the rural communities lack potable water and has to travel many miles to search for water in nearby polluted streams for domestic uses. (Table I) At Agakwe, in Tiv Land, Edoma land in Benue state of Nigeria, pipe borne water was found (Survey done by researchers with CARUDEP, JOS, 2005) while in Bauchi, an all the local Governments in the rural areas of Tafawa Balewa. Ganjuwa etc depend on wells that dry up in the dry season. The same holds true for many communities in the northern parts of Cameroon, Chad, Sudan, Central African Republic and Niger. The communities lack basic hygiene, sanitation and water (Table 2.) UNICEF (1993, 2009) acknowledged that the lack of universal access to health, education and water services for the world's poorest people is a big obstacle to the global targets for sustainable development (UNEP, 2002) Unfortunately, this obstacle remains and it is uncertain if the strategies on ground can generate sustainability in any way.This is because, poor people in semi-urban and some rural communities in subsharan africa still pay a disproportionate share of their meagre incomes for water services that is irregular, inconvenient and often suspicious in quality.A survey done in some villages in Cameroon shows just how potable water is rare. (Table1).Paradoxically, so much attention, the world over has been placed on water pollution and sanitation programmes with huge expenditures but the impact, however, remains questionable. Perhaps to attempt to explain this short falls could be that most of the strategies used to solve these problems are in themselves not sustainable.

Community Name	No. adult male	No. of adult women	Children 1-12	Youth 12-25	No. Latrines	No houses	Water point
GarinAbare	400	800	1500	700	120	300	5
GIKAR	1000	2250	4000	700	21	250	3
KWABLANG	250	350	1000		32	100	19
Barkaya	500	600	3500	1500	32	100	6
DUKKUN Dindima	25	36	59	18	32	13	19
Turiya	25	38	88	15	15	11	9
Nassarawa	35	42	81	19	26	29	5
Dindima	25	36	56	19	38	13	2
Fumbinare	42	59	102	54	49	12	5
G/Total	2302	4211	10386	1625	365	828	73

*Done in collaboration with Development Exchange Centre (DEC), an NGO which provides / assist these communities with development projects in 2005

Table 1. Baseline survey on water and sanitation facilities from communities in Bauchi state, Nigeria (The survey indicates limited safe water and sanitation among the rural area in Bauchi-Nigeria)

Illnesses	Adults No%	Children
Scabies	24 (2.9%)	32 (3.8%)
Skin sepsis	16 (1.9%)	20 (2.4%)
Yaws	4 (0.5%)	10 (1.2%)
Lice	152 (18.3%)	250(30.0%)
Trachoma	16(1.9%)	18(2.2%)
Conjuntivitis	65(7.8%)	133(16.0%)
Bacillary dysentery	106(12.7%)	198(23.8%)
Salmonellosis	16(1.9%)	28(3.4%)
Diarrhoea	14(1.7%)	108(13.0%)
Ascariasis	10(1.2%)	18(2.2%)
Paratyphoid Fever	20(2.4%)	34(4.1%)
Worms	4 (0.5%)	24(2.9%)
Stomach ache	-	103(2.9%)
Malaria	70(8.4%)	6(0.7%)
	517(53.7)	986(117.4%)

Table 2. Frequency distribution of common illnesses found in Bulli Village, 10km away from the university Community in bauchi, Nigeria. (The results in this table indicates high frequency of water borne diseases in the study area)

Providing pipe borne water to communities, is laudable but when the low income earning communities cannot cope with maintenance cost and the robustness of the technologies in place, then it becomes a major problem.UNICEF (1993) reported that in the 1980s, some 10 million dollars was spent yearly in the developing countries on high technology to improve services to people who already had water and sanitation predominantly in the cities. Only a fraction (20%) was reluctantly spared on low-cost appropriate technology for the underserved majority of people in peri-urban areas (UNICEF, 1993).The high cost of treating water and its attending high energy input is prohibitive to most industries and factories in developing countries and as such release untreated wastewater into neighbouring streams, thereby polluting many fresh water bodies. These sources of water pollution includes heavy metals, halogenated hydrocarbons, dioxins, organochlorines such as DDT which do not easily break down under natural processes and tend to accumulate in biological food chain.The popular treatment system of water in sub saharan africa is the sedimentation, coagulation, disinfection (chlorination), filtration. Undoubtedly, this has generated potable water but, however, the final water products remain unaffordable by 70% of ther populace (Schultz et al., 1983; Yongabi et al., 2010) Reports also suggest chlorine resistant organisms such as, cryptosporidium oocysts, strains of salmonella sp, aeromonas, entamoeba cyst, mycobacterium sp, escherichia Coli. 0157:47 and host of others (Madore et al., 1987;Yongabi et al., 2011).Chlorine has been noted as a potential carcinogen forming compounds such as tetrachloromethane (TCM) which also produces hormonal analoque that may interfere with male fertility. Aluminium sulphate (Alum), the widely used water coagulant thus generate acidic water, unsafe for pregnant women and causes predementia in some people (loss of memory) While all these defects exists, mankind has been endowed with indigenous knowledge and has been using it to survive proceeding the advent of all these technologies. There is a need to revisit our roots, study this indigenous system and improve on them. Exploring and exploiting the potentials of natural materials such as plants and sand to bring about cheap clean water in a more ecological friendly manner are the thrust of this work. This may have great lessons for ecological sustainability now and centuries to come.

1.2 Aims / objectives

The ultimate purpose of this chapter is to report results of our research on a water pollution management technology that is low-tech, cheap and above all ecologically friendly. The specific objectives of this study, therefore, are:to report the results of analysis of the pathogen level of polluted water from refinery, food and confectionery processing industry in Nigeria and Cameroon, stagnant pond water where people fetch water for household chores and for irrigation at. To carry out a survey / inventory on problems of clean water and indigenous knowledge on how communities treat their water in Nigeria and Cameroon. To use the collected knowledge and screen these plant materials and their extract for their coagulation/ disinfection activities in vitro using polluted water samples.To test their potential antimicrobial activity on isolates from polluted water samples and, to generate clean water using a constructed integrated biocoagulant - sand filter system and other geological-materials.

2. Brief overview of interdisciplinary importance, dangers of water and existing gaps in water pollution management

2.1 The necessity of water as a consumable product in all the aspects of life

The role of water in life as a whole cannot be over emphasised as this universal solvent is the basis of life after air. What a life without water? Evolutionary, biologists hold strongly that life began in water and therefore explains why human use water at times for rituals.Water is a prime necessity for life, it forms the basis for a balanced diet without which digestion cannot function well. It is a lubricant for biological processes such as excretion and major glands secretion are usually in water form.It acts as a cushion preventing crushing in internal structures, example synovial fluid.

To the agriculturalist, crops cannot grow without water, therefore, it is needed for germination, that is probably why in dry areas irrigation is used to shunt this adversity.Water is used for laundry, domestication such as cooking and washing of utensils. This vital community is the basis for electricity in which case more important than electricity. It is the source for hydroelectricity which is the backbone of all industries and factories.Additionally, water is used in agro-industry for washing, media for dissolution, production of diary products, beverages etc.To the engineer, water is used as a cooling agent, lubricant and for building and construction.In addition, water is a useful transport source: Navigation. Water is a habitat for fish and minerals such as petroleum. Fish being used as sources of protein for man and petroleum and other minerals used as fuels.

Water serves as a touristic site, for example the Kribi beach in Cameroon and the Gubi dam in Bauchi state. In the wise, a source of revenue for the Government. Indeed, highlighting the use of water could only tatamounts to an infinite list. Taking water as previously mentioned is a source of a balanced diet, it contains vital minerals such as Manessium, Ca, Fe, Cu, Zu, F, No_3, So_4 etc) for the International Standard for drinking water (WHO, 1984).Therefore, good water should actually posses these minerals. Good water should be colourless, odourless and free from any toxic elements.Some toxic substances in drinking water could include Pb, Se, As, Cr and CN. According to World Health Organization, 1958, showed that a 0.001 4p.m s the maximum concentrations allowable, exceeding this level is pollution (see As cited above, toxic elements could be consumed from water that could lead to cancer. Water, even though has many uses serves as a breeding ground for some vectors of man's parasitic diseases, for example. Malaria and schistosomiasis. Rain water in excess could cause flood and hence heavy economic losses. Besides, this could also lead to erosion which inflicts heavy pains to Agriculture. Finally, sea accidents lead to loss of lives too. The overwhelming indispensability of water as a primordial stuff for all the arms of Economy has become a hot topic for discussion by many state governors and their administrations in africa.If the State governments are not supplying boreholes and other rural water Sources, she is either trying to solve flood problems or some other hazard cause by rain storm.From analysis of all budgets speeches since 1982 to date, it is interesting to note that water has been placed as a top priority amongst other projects yet little is achieved. Water which is safe for drinking must be free of pathogenic organisms, toxic substances and an excess of minerals and organic debris. It must be colourless, tasteless and odourless in order to be attractive to consumers and preferably cool.Water is the basis of life. About 75% of the body weight is made of water. In developing countries 15 million infants die every year due to contaminated drinking water, poor hygiene and malnutrition. About 80% of illness in developing countries are directly connected with contaminated drinking water (WHO).The Provision of water supply near by for consumers and sufficient for their daily needs will help greatly in decreasing the incidence of skin diseases and eye infections and also reduce diarrhea diseases and most worm infections, particularly if the water is of good quality bacteriological. However, major improvements in health conditions through provision of sufficient safe water can only be achieved through domestic hygiene and proper methods of water purification (Yongabi et al., 2010).Oyawaye et al (2000) in their study of water sources for three years in Bauchi, noted elevated levels of nitrates (33.3mgkg) in ground water sources in the dry season.Higher nitrate values for treated and untreated waters still remained high in the rainy season but within acceptable limits.Excess Nitrates in water has been linked to methemoglobinemia (bleu babies) Many infant deaths in Africa and particularly sub Saharan Africa are mainly attributed to dysentery and diarrhoea of undefined sources which may be due to nitrates in water. Similarly, literature elsewhere has evidence that implicates N- Nitrosamines in the inccidence of carcinogenesis. The density of Microbial isolates has been reported to be inversely proportional to the level of residual chlorine from 1.0mgk to less then 0.2mgk. Residual chlorine also reduces steadily from point of application to point of collection. Twenty three bacterial genera belonging to groups of coliform, faecal coliform and Staphylococcus spp were isolated at various stages (Yongabi et al., 2011).

2.2 General methods of water pollution management
2.2.1 Application of chlorine, halogens and alum in water treatment
Chlorine is widely applied to disinfect water. For instance, a well or a tank containing 1000 litres of relatively clean water 2g of chlorine is added and if organic matter is present or one is doubting the purity you add 4g of chlorine. Then thoroughly mix it into the water and allow standing for at least 30 minutes before using it. However if water is highly turbid i.e containing a lot of sediments, alum is first added to make the sediments settle at the base. The water is then drain into another tank before chlorinating. The amount of Alum required treating 1000 litres of relatively clean water is 56g while for sufficient safe water for a community but then it requires highly skilled technicians who can measure and control the chlorine and alum dosage. This knowledge is lacking in the rural areas.Other halogens such as bromine and iodine are also applied in water treatment. The set backs have been discussed in recent publications (Yongabi et al., 2010 and Yongabi et al., 2011).

2.2.2 Sand filter
This method of purifying water has been known right from time immemorial. Over thousands of years now clean water have been obtained from river beds when dug. As water falls or flows over the river bed it percolates through the sand grains where the disease-causing organism filter out. Clean Sharp River sand is obtained and thoroughly washed; gravels are also obtained and washed. Tow clean containers are used for the construction of the sand filter. The container for the filter and storage could be made out of metal plastic or traditional clay. A hole is made two-thirds of the way up the filter container and hose with blocked base and perforated will be fixed at the opening into the drum. The gravel is then placed over it to a height of 7.5cm and the sand is placed above it to a height just below the hose fitting. The filter is then thoroughly flushed out with clean water for a week to allow for formation of biofilm. The attributes and set backs of the sand filter in terms of cost, installation, management and efficacy and the need to intergrate it with plant coagulants have been reported (Yongabi et al., 2010).

2.2.3 Water treatment with plants: The case of moringa oleifera and water hyacinth
The seed pods of Moringa oleifera have been used for water treatment. After shelving, the seeds are crushed, seized (3.5mm mesh) using traditional techniques employed in the production of maize flour. Approximately 50-150mg of the ground seed will be needed to

treat a litre of river water, depending on the quantity of suspended matter.Normally, a small amount of clean water is then mixed with the crushed seed to form a paste.The crushed seed powder when added to water, yields water soluble proteins that posses a net positive charge.The coagulant/flocculant characteristics of seed is linked to a series of low molecular weight cationic protein. Dose of Moringa oleifera seeds depends much on the tubidity of the water in question. Generally, 75-25mg/l (.75 - 2.5g) has been empolyed.For a turbidity of 400NTU,).5g of Moringa oleifera powder is used for a litre of turbid water.Extensive studies have been done on the applications of Moringa oleifera in water treatment. Other plants used in water treatment include, cactus, water hyacinth and syntrochnus potatorum which are reported to remove turbidity and heavy metals from water (Bina, 1991; Yongabi et al., 2010) The need to catalogue such useful natural materials in Africa needs intensification.

3. Materials and methods

3.1 Study area
The study was conducted in Bauchi State (at Abubakar Tafawa Balewa University) Nigeria and Bamenda, Cameroon 2010 to 2011. These two countries are located in sub shaharan Africa with the same climatic conditions and similar traditions and problems. These Natural plant materials collected from these focussed more on other plants rather than Moringa oleifera which has been extensively reported in literature.

3.2.1 Materials used
MacCathney and bijore bottles were purchased from supplies of Hospital and laboratory materials from bauchi metropolis. They were washed repeatedly using detergent and rinsed in clean water and then sterilised by autoclaving alongside with all glasswares used for the study. Autoclaving was done at 121°c for 15 minutes.
A number of agars: Nutrient agar (oxoid Ltd), MacConkey Eosine Methylene blue, potato Dextrose agars (Oxoid) Ltd) were obtained from the University Zeri Research laboratory and Phytobiotechnology Research laboratory and School of Chemical engineering , The University of Adelaide, South Australia.

3.2.2 Special equipment and apparati
Some of the equipment and apparatuses used for the study include spectrophotometer (Phips) Vu/Vis, Pve, unicam sp 6-450) incubator (Jouan) bench. Certrifuge (Mistrail 1000) weighing balance (meter am100) and soxhlet apparatus (Galler kamp).

3.2.3 Chemicals and reagents
Nutient Agar were obtained from biotech laboratories survey, UK, Ferric chloride, potassium hydroxide, copper acetate, lead acetate, bismuth nitrate sodium chloride, chloroform, diethylether, ethanol were purchased from Britiest Drug Houses (BDH). Chemicals Ltd, poole England.
Ammonia solution potassium tartrate picric acid, fhling solutions were obtained from Ma nd B Ltd England.
All other regents an chemicals used were of analytical grade obtained from reputable scientific and chemical companies. All solutions were prepared in distilled water, redistilled from pyrex apparatus.

Carica Papaya Plant
seeds used as a phydisinfectant and coagulant in rural Cameroon

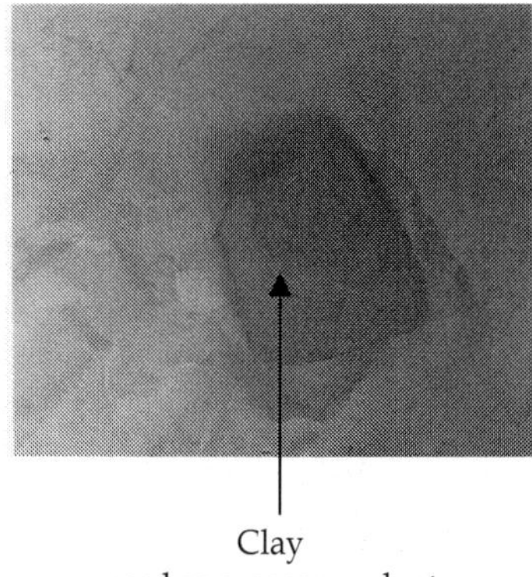

Clay
used as a geocoagulant

Pieces of Alum   Garcinia Kola seeds

Alum is a synthetic coagulant   used as a phytocoagulant and phytodisinfectant,
widely used, people have to buy.   locally available in the communities in Africa

Jatropha Curcas Plant

Seeds used as a phytocoagulant, locally available in Subsaharan Africa, particularly Cameroon

3.2.4 Polluted water sample collection

Refinery wastewater was collected from other Kaduna Petroleum oil refinery. This is located in kaduna Town in Kaduna State in the northern parts of Nigeria and SONARA oil in Cameroon. The crude oil is fractionated and fuel produced amongst other, bye products. The refinery wastewater is usually discharged untreated into river Kaduna. Ten litres of the wastewater was collected with the assistance of students undertaking their internship at the refinery in 2005-2011.Wastewater from the NASCO Company Ltd in Jos, plateau State of Nigeria was also collected. The NASCO household in Jos produces a number of confectioneries including, biscuits, conflakes etc and then Nasco soaps, detergents etc. Jos is located in the North Central region of Nigeria and has a teeming population. The wastewater is usually discharged into the neighbouring streams and brooks. Ten litres of the wastewater was collected with the acid of students on internship.Lastly, dirty (turbid) pond water was collected from a stagnant pond located at the western part of the Abubakar Tafawa Balewa University Campus. The stagnant water is used by the local people around for irrigation of Crops within the vicinity, other activities include fishing and washing of clothes as well as at times swimming. Ten litres of the sample was collected for laboratory studies.

3.3 Microbiology analyses

One ml of each of the samples was diluted in 9ml of sterile distilled water and serially diluted up to 10-5 dilution and plated in triplicates on Nutrient agar for total heterotrophic bacterial counts, MacConkeyagar and Eosine Methylene blue agars for Total Coliform and *E. coli* counts respectively while on potato dextrose agar for fungal counts.Incubation was done at 37°C for 24 hours for bacterial counts and at 25°C for fungal counts. Discret colonies on each plates were counted on each plate and average of three plates taken.The presence of colonies on Eosine methylene blue agar indicated probable identify for typical coliform colonies, Gram stained portions of the colonies showed gram negative roads with absence of spores as a further elucidation of the Micromorphology of coliform. Colonies on EMB that appeared as Metallic greenish sheen confirmed the presence of Escherichia Coli.

3.3.1 Collection and Identification of plants

The leaves and *Abus precatorius* were collected from Shere hills, Jos Nigeria. This plant sample was authenticated by plant taxonomists at the Federal College of forestery Jos,

Nigeria. They were dried and pounded into well labeled, clean air tight. Containers and stored until required.

3.3.2 Sources of test organisms

Clinical isolated of Esehericaia coli, staphyococcus aureus, Salmonella paratyphi and Candida albican were isolated from polluted streams in Bamenda, Cameroon.

3.3.3 Methods

Preparation of plants for biological test

The dried and pulverized samples were then extracted using ethanol, diethyl and water. This was done in increasing polarity. Most traditional healers sometimes use palm wine, which contains ethanil as their solvent. This gives additional reason for preferring ethanol to methanol. Deithyl ether is one of the best solvent for antimicrobial activities (nastro et al, 2000). Water is a universal solvent.

3.3.4 Aqueous extraction

10g of powdered sample was weighed on a Mettle balance. It was then put in separate clean and sterile conical flasks containing 200ml of cold sterile.

3.3.5 Biological assay

Preparation of dilutions of the extracts

The concentrations of various crude extracts were made in sterile distilled water and for dicthylether extrac) the concentrations prepared in those solvents were 50ml, 100mg/ml and 150mg/ml.

3.3.6 Purification of bacterial isolates

The stock cultures of the bacterial isolates were subculture unto nutrient agar, blood agar and agar and macconkey agar to produce discrete colonies and incubated for 24 hours at 37°C. The plates were examined for purity and specific biochemical test were carried out to confirm the identity of the different isolates according to methods described by (Baker et al, 1980).

3.3.7 Test for bacterial suspensions

Preparation of fresh plates of the test bacteria was made from isolated stocks stored on agar slants. By the use of a stride wire loop, colonies of fresh cultures were picked and suspended in 20ml of nutrient broth in different sterile universal bottles. The centrifugation bottles were done in MSE refrigerating centrifuge at 1000m rpm for 30 minutes in virology department of N.V.R.I Vom. The supernatant was discarded. The organisms were again resuspended using equal volumes of sterile normal saline. The concentrations of the organism were obtained by comparison with 10 standard opacity bottles (Macfarland's Naphelometry) method of opacity, which contained various amounts of barium sulphate in 1% sulphereric acid (N/36). Most of the tubes corresponded to 10-4 which was very turbid. The organisms were then diluted down to 10-6 s then one loopted equivalent of 0.02ml from each of the bottles 10-4, 10-5 and 10-6 was plated out on three different petridishes containing nutrient agar and incubator over-right at 37°C to determine population density of the test organism. Member of colony forming units per millilitre was obtained as follows, since for example 10^{-6} deliration had 25 colonies. Colonies on the second day 25 x 50 x 10^{-6} + 1.25 x 10^{-9} C.FU/ml. 1ml of the 10-6 dilution of various bacteria was used in flooding nutrient agar plates in the agar diffusion method of invitro sensitivity test.

3.3.8 Preparation of media

2.3g of nutrient agar was dissolved in 100ml of distilled water and heated slowly while shaking until the solution become clear and yellow in colour. The nutrient agar was cool to about 47°C and become semi-solid state. This is to facilitate the diffusion of large molecules of the crude extract as compound or standard or processed and purified antibiotics with small and readily diffusable molecules.

3.4.1 Agar gel diffusion test (punch-hole method)

The plates of nutrient agar were seeded in duplicates with 1.0ml of 10^{-6} dilution of the test bacteria. The plates were then swirled to allow the inoculum to spread on the excess was discarded in a disinfectant jar. The plates were allowed on the bench for 5 minute is and they were dired in the incubator for 1 hour at 37°C.

Using a sterile cork borer four well were bored at equal distances around for plate. The 5th well was made in the middle. The bottoms of the wells were sealed with one drop each of sterile nutrient agar before the extracts were puts.

The prepared concentrations the extracts were put into the wells. Sterile distilled water was put in the 5th well to serve as negative control for aqueous and ethanolic extracts while dimethysulfoxide ws used as negative control for diethylether. Gentamycin was used as a positive control in the 4th well. After allowing on the the bench for 1 hour, for diffusion of the extracts, the plates were incubated at 37°C for one day. The plates were examined the next day to concentrations of the extracts on the test bacteria.

The zones of inhibitions were measured using a ruler in millimeters and the average of the two readings was taken to be the zone of inhibition of the bacterial species in a particular concentration.

3.4.2 Minimum Inhibitory Concentration (MIC)

This was determined using broth dilution technique (Puyelde, 1956). Freshly prepared broth in sterile Bijou bottles was used. Two sets of six Bijou bottles were used for each test. 1 ml of sterile nutrient broth was put in Bijo bottle number 1 to 6.1ml 200mg/ml was added to Bijou bottle number one. The extract in the bottle on was therefore diluted 1:2. It was properly mixed and 1ml was transferred to bottle number two which was diluted 1:4 and this was continued until the 5th bottle from which one ml was discarded. Bottle number six contained only sterile nutrient broth to serve as negative control.A loopful of 10^{-6} dilution of bacteria suspension with microbial load of 1.25 x 10^9 C.F.U/ml was then added to all the six bottles. This entire procedure was done for all the organisms that were susceptible to the various extracts. The bottles were thoroughly mixed by gentle shaking and incubation for 24 hours at 37°C. The bottles were observed for turbidity after incubation visually by comparing with the control.Cultures from incubated bottles were subcultured onto fresh nutrient agar plates. The inoculated were incubated at 37°C for 24 hours. The plates were examined for growth indicated bacteriocidal effect of the concentration of the extract used. Plates showing light growth were taken to have bacteriostatic effect, while those showing moderate or heavy growth were taken to have no inhibitory effect on the bacteria (Puyuelde, 1986).

3.4.3 PH analysis

The PH of the raw and treated wastewater samples was tested using a combi-9 test trip (a standard strip for routine urinary biochemical analysis). A fresh strip each was dipped into each of the samples and after sixty seconds, the colour change noticed was compared with a range of colour standards and when the colour of the strip Matched any of the colour standards, the PH label was directly read off.
(Photo field solar weighing balance and Combi-9 Ph strip).

3.4.4 Turbidity evaluation

A subjective visual observation was done. The presence of colloidal suspended matter was noted in the untreated samples while their absence noted in the treated samples Floc formation and lack of floc formation was also observed as a distinct evidence of coagulation for the treated samples.The presence of odour and absences was also noted by suing the nose. The use of the sight and small senses were highly exploited.

3.4.5 Plant sample selection and collection

The plant coagulants used in this study were selected based on a survey of their local use in water purification by the indigenous people in sub-saharan africa (Yongabi K. A, 2004, www.biotech.kth.se/iobb/new/kenneth04.doc) Moringa Oleifera (Lam) seeds have been used by a rural Nigeria for water treatment and Literature elsewhere abound (Fuglie, 1999, Folkland et al, 2000,) The dried seed of Moringa Oleifera were harvested from Bauchi State, Nigeria. Seeds of Garcinia Kola, Hibiscus sabdariffa and Carica papaya were collected from Enugu in Nigeria and Bamenda, Cameroon (Photo).

3.4.6 Plant processing

The seed pods were harvested and stored in Khaki envelopes, They were deshelled (specifically M Oleifera Garcinia Kola) while the seed of Carica papaya were scoped out from riped fruits as well as Hibiscus sbadafrifa seeds were purchased from the market at Mdulawal Market in Bauchi Metropolis.

3.4.7 Coagulation studies

Graded weight (0.5g to 5g) of the pulverized plant Materials each and Alum, Hydrogen peroxide, were each added to 200mls of each of the wastewater samples in 250ml capacity beakers.
Increased weights in grams from 0.5g to 5.0g of each of the plant material was mixed in a small quantity of turbid water for form a paste and then mixed carefully with the water samples in the beakers.
The same procedure was done for Alum and a turbid water sample in a beaker (200mls. was allowed to stand in a beaker for 24 hours as controls). The Coagulative effects and change in total bacterial counts, PH, visual clarity amongts other parameters were evaluated.

3.5.1 Cold extraction (Buck extraction)

A cold Methanol and aqueous Extraction was then carried out on 50 grams each of Hibiscus sabdarigga seed and *Carica papaya* seed powders except for Moringa Oleifera. 50 grams of each of the powders was steeped in 250mls each of methanol and water for 24 hours. Gravity filtration was carried out using whatman filter paper N° 13 and solvent evaporated at room temperature.

3.5.2 A cold sequential extraction of *Moringa oleifera* and *Garcina kola* seeds

A cold Sequential solvent Extraction was carried out on Moringa oleifera seed powder using n-hexane, Dichloromethane Methanol and water in that order. The purpose of this was to exploit the polarity effect of the solvent on the possible isolation of the active portion from the plant material 50grams of the pulverized seed (pulverised using a pestle and mortar) was steeped in 250ml of n-hexane left for 24hours, filter off using gravity filtration using whatman filter paper No 13, The plant residue was dried in the sun and used for the next solvent and the order maintained for all the other solvents.

The extracts were left in the open for 2 weeks for the solvent to evaporate. The extracts were now used for antibacterial bioassay.

3.5.3 Antibacterial assay (agar diffusion method)

The bacterial isolates were re-cultured in peptone water for 18 hours and 0.3ml of each of the bacterial suspension was mixed aseptically with 15ml nutient agar (oxoid) in sterile petri plates and allowed to solidify. A stainless steel borer of 6mm diameter was used to punch wells into the agar and each well was filled with 0.1ml of 2% extract, and with oil and of sterile distilled water, H202 and Alum as controls.

3.5.4 Phytochemical screening

The phytochemical screening of the powdered extracts obtained from the leaves of Abrus precatorius were carried out using standard qualitative procedures (Trease and Evans 1989, Sofowora 1986).

3.5.5 Test for alkaloids

Two grams of plants materials thoroughly grounded was treated in a test tube with 25ml of 1% Ad for 15min in a water bath. The suspension was filtrated in a test tube and the filtration was divided in two pants A and b.

To filtrate A, five drops of Dragendorff reagent were added. The formation of a precipitate indicated the presence of alkaloids.

3.5.6 Test for flavonoids

i.   Well ground plant material (1g) was extracted with water (10ml) and methanol (5ml) and filtered. Few magnesium turnings were added to 3ml of filtrate and concentrated added dropwise (cyanidine reaction). Developments of colour indicate the presence of flavoniods a red colour and flavonones give a pink colour.

ii. To 1ml of the extract 1ml of Naoh was added. The formation of a golden yellow precipitate indicated the presence of flavoniods.

3.5.7 Test for cardiac glycosides (salkowsi test)

0.5g of extract was added to 2ml of chloroform and after mixing, 2ml of H2 so were carefully added to from a lower layer. Reddhis brown colour at the interface indicates the presence of a steroidal ring i.e. glycogen portion of the cardiac glycoside.

3.5.8 Test for anthraquinones

Anthraquinones are a sunset of anthranoids. For the specific test an ether chloroform maceration (1gin 5ml of CHd3 and 5ml of ethered was filtered and 1ml of 10% NaoH solution. A red quinones. A weak coloration was assigned a +, while a strong caloration a +++.

3.5.9 Test for steroids

Powdered plant material (1g) was covered with either and shaken occasionally for 2 hours. The solution was filtered and decanted. 1ml of the solution was put on porcelain plate to evaporate. A drop of conc. H2So4 was added and stirred orange colorition was positive indication.

3.6.1 Test for saponins

Well-grounded plant material (1g) in water (15ml) in a test tube was heated on water bath for 5 minutes. The solution was filtered and left to cool to room temperature. The filtrate (10m) in 16 x 160mm test tube was shaken for 10 second and the height of honeycomb troth, which persisted, was measured. Froth higher than 1cm confirms the presence of saponins.

3.6.2 Test for tannins

10ml of water were added to 5g of extract ad the mixture was stirred and filtered. To 2ml of the filtered. To 2ml of the filtrate few drops of 0.1% fecl$_3$ solution and the development of precipitate was observed. A blue-black, green precipitate indicates the presence of tannins.

3.6.3 Test for carbohydrate

5g of the powder sample was boiled in 1oml-distilled water on hot plate for 5 minutes and filtered whole hot. The filtrated was used for the following tests.

i.   **Molisih test**

To 3.0ml of the 2 filtrate was added 3 drops of molisch reagents then carefully run 3.oml conc, H_2SO_4 without shaking. The interphase formed was then observed for purple.

ii.   **Benedicts test**

3 drops of the filtrate was added to 2.0ml of benedict reagent and placed on a hot plate for 5 minutes to observe the formation of brick red precipitate

3.6.4 Balsam test

To 3 drops of alcoholic ferric chloride was added to 2.0ml of extract then warm a dark green coloration if formed with balsam. To 2.0ml of the extract were added few drops of potassium permanganate. The solutions was then warmed on hot plate and observe for benzaldehyde or almond adour.

This was carried out in duplicate, and each set up was incubated at 37°C for 24 hours and the diameter of zone of inhibition in mm was recorded using a vernier caliper.

3.6.5 Phytochemical screening
3.6.6 Test for alkaloid

Twenty (20) mg of each of the extract was placed into a test tube, 1ml of distilled water and 2 drops of 1% HCL were added and the solution was warmed gently in a waterbath to effect complete dissolution of the extract. A stream of dragendorff's reagent was added tot he solution from a test tube.

3.6.7 Test for glycosides

A ml (10 of each of the extract solution was placed in a test tube and a drop each of 2% 3.5 dinitrobenzois acid in Methanol and 5% OH in water was added.

3.6.8 Test for tannins

A ml (1) of each of the extract was placed in a test tube and a stream of 5% $FeCl_3$ solution was added.

3.6.9 Test for flavonoids

Twenty (20) mg of each of the extract was dissolved in 2ml ethanol in a test tube, a small size spatula full of zinc powder was added and a few drops of HCL was then added.

3.7.1 Test for soluble carbohydrates

Twenty (20) mg of each of the extracts was dissolved in 1ml distilled water and 2 drops of 5% L-nahthol solution in methanol added in a test tube. While holding the tube at an angle, a stream of cone' H_2so4 was added to it.

3.7.2 Test for saponin

Twenty (20)mg of each of the extracts was dissolved in 1ml of distilled water and 2 drops of 1% Hcl was then heated gently on a water bath.

3.8 Construction of a sand filter

The design of a sand filter using two 200 litres plastic drums. The drum is cleaned out and hole is made two thirds of the way up so that an outlet pipe can be filtered. Depending on the size of the nipple, the hole is made. The water-collecting pipe is made with of hose piping. This is connected to the outlet pipe by a short hose piping. A number of saw cuts of drilled hole are made in hose piping ring and this is laid down on the bottom of the drum. The second drum is constructed for a storage drum. First a hole is made at the same level as that on the first drum and an appropriate nipple is fitted. A connecting hose is fix from the filter to the hole on the storage drum. Another hole is made at the other side and at the bottom of the drum at a height of about 7cm from the base and a water collecting pipe if fitted such that it is long enough to be dipped at the top. In other cases a tap could be filled for collecting water, but this can easily become loose as a result of constant opening and closing, so the hose is more preferable.

For setting up of the filter, clean sharp river sand of different sizes are obtained form a riverbed and sieved out. Gravels and coal of the correct sizes are also obtained and thoroughly washed with clean water; the sand is also thoroughly washed too kept in a place safe from dirt and dust.

3.8.1 Sand/gravel filter

This was constructed using sand and gravel only as media. The gravels are first placed at the bottom to a height of 75mm (7.5cm), this is followed by a layer of coarse sand to a height about 100mm. The last layer of fine sand is placed to a height just below the level of outlet pipe. This arrangement is made so that even if the tap is left on, the water drains out of the filter, a small layer of water remains above the sand. The sand must never be allowed to go dry, otherwise the biologically active ingredients in the sand which are important to the purification process, will die out. Both drums should have a lid to cover the drum, and this is made with a sieve or strainer for water with allot of sediments. The sieve should be covered too.

When the filter has been completed, it must be thoroughly flushed through with clean water to further remove any dirt present. Once this is completed, a daily routine of adding raw water is maintained for a week or more so that the filter skin can form before usage begins.The design of the working components of the filter using a 200 litre drum should provide at least 624 litre of water per day. The yield rate will be controlled or regulated to 0.4 litres/minutes because the rate of flow of the filtered water to be slow to ensure satisfactory performance. However, if the raw water to be filtered has a bad odour, taste or colour, player of coal can be introduced between the sand and the gravel layers to control the situation.The procedure outlined about is for a 200 litre drum but the same technique can be used for a brick built container, metal drum, or clay pots.Below are some models of sand filter using different media and each had the calculation of the yield rate as a guide for other types of containers.

4. Results highlights

The picture below shows the turdity clearance level with various treatments including untreated storm water left as control

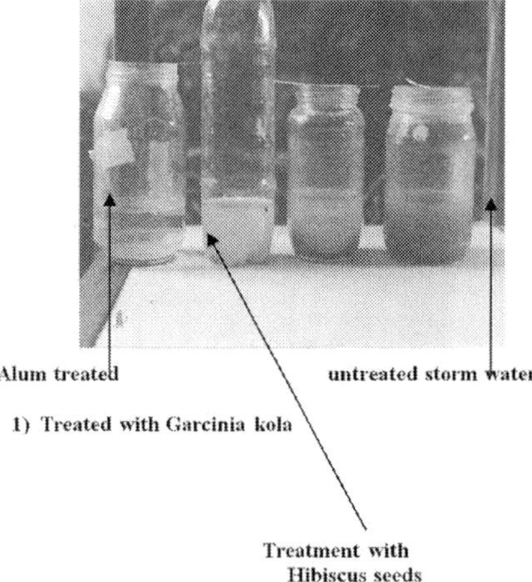

Storm water being treated with Alum in 15 minutes, water is appears clear.1 storm water treated with garcinia kola seeds, particles settles comparable to Alum.The third treatment container from left is storm water treated with Hibiscus seeds, particles settle but not as clear as Alum and Garcinia.The fourth treatment container from left is untreated storm water left as control, less settlement of particles.

The findings are presented in the following tables.Data in table 1a shows the pH and bacterial counts of foul wastewater from refinery, in the untreated wasyewater, the total bacterial and fungal counts were high with a strog foul odour.Pseudomonas was also isolated in the untreated wastewater. In Table 1 b, after treatment with plant materials, the total microbial counts dropped significantly to tolerable levels, the pH was stabilized while odour was no longer perceived.pseudomonas spp was no more isolated.However, the various degree of treatment varies with the different plant materials applied.Moringa oleifera, Garcinia kola and Carica papaya exhibited the best results.

Type of Treatment	Colour	PH	Smell	Appearance	THBC Cfulml	Coliforms Cfulml	E. Coli cfulml	TFC
Untreated OVH	Colourless	6.6	Engine Oil, crude Oil smell	Clear	560	Nil	Nil	315
Untreated DFH	Brownish	7.05	strong Engine oil Smell	Turbid	300	nil	nil	6140

* OVH ---> Overhead fraction or foul wastewater * Pseudomonas spp isolated
* DFH ---> Desalter foul water

Table 3. (a) Effect of plant seed powders and Alum on oil Refinery wastewater from Kaduna State, Nigeria (Significant reduction in turbidity and microbial load using plant coagulants and disinfectants as indicated in the table)

Type of Treatment	Colour	PH	Smell	Appearance	THBC Cfulml	Coliforms Cfulml	Coli cfulml	TFC
(b)								
Untreated OVH	Colouless	6.6	Engine oil, crude oil smell	Clear	560	Nil	Nil	315
Treatment with Moringa Oleifera seed	Very colouless	7.0	odour absent	very clear	36	Nil	Nil	100
Treatment with Garcinia Kola seed	Very Colourless	7.0	odour absent	Clear	70	Nil	Nil	173
Treatment with Carica Papaya seeds	Colourless	7.0	odour absent (papaya odour)	Clear	62	Nil	Nil	87

Treatment with Hibiscus sabdariffa seeds	Colourless	5.0	odour (faint)	Clear	133	Nil	Nil	113
Treatment with Alum	Very Colourles	5.0	Odour (faint)	Very clear	313	Nil	Nil	6140

(c)

Untreated DFH	Brownish	7.05	strong engine oil smell	Turbid	300	Nil	Nil	6140
Treatment with Moringa oleifera seed	Very clear colourless	7.0	odour absent	Clear	96	Nil	Nil	89
Treatment with Garcinia Kola Seeds	Clear colourless	7.0	no tment with	Clear	89	Nil	Nil	125
Alum	Clear (very)	5.0	odour persist faintly	Clear	122	Nil	Nil	168

In table 4 below, the results indicated that plant materials exhibited great disinfection potentials on grey water (detergent based water) when compared to alum.

Types of treatments	THBC	Coliforms	E Coli
	Cfulml	Cfulml	Cfulml
Untreated waste water sample	2,200	2 300	1,900
Untreated waste water sample left on bench and analysed	2,120	2,224	1,892
Alum treated sample	600	1.070	
Moringa Oleifera treated sample	320	520	343
Jatropha Curcas treated sample	770	890	729
Garcinia Kola treated sample	700	675	521
Carica papaya treated sample	697	682	575
Persea americana treated sample	800	760	690
Hibiscus sabdarrifa treated sample	600	800	

* Wastewater normally stored for a week and then disposed 5g of powders of plant seeds used.

Table 4. Effects of Plant seed powders and alum on grey water detergent based water from Nasco Factory Jos, Nigeria

Table 5, the data shows that the physicochemical properties of the detergent based water such as turbidity and pH was significantly reduced when treated with the plant based coagulants when compared with the untreated wastewater sample.

Treatment Material	Turbidity assessment	PH	Remarks
Untreated wastewater sample	Very turbid, foamy bluish	6.5	Odour intense turbidity remains the same
Untreated wastewater sample left on bench and analysed	Remains turbid, foamy.	6.5	Odour intense turbidity the same
Alum treated sample	Fast precipitation, very clear, slight odour	5.0	___
Moringa Oleifera treated sample	Flocs formed, odour totally removed	7.0	Odour removal colour removal protein positive.
Jatropha Curcas treated sample	Flocs formed settled at bottom	6.0	___
Garcinia Kola treated sample	Clear, flocs formed with a suspended pellicle	6.0	Second stage treatment clearer, odour off, after proper filtration
Carica papaya treated sample	________	6.0	________
Persea americana treated sample	clear, no odour flocs settled at the bottom	6.0	a second stage treatment was better.
Hibiscus sabdariffa seeds treated sample	Flocs settled at the bottom must slight odour	6.0	Protein positive

Table 5. Physicochemical properties of treated and untreated detergent based wastewater from food/detergent factory

In table 6, the data shows that the plant seed powders demonstrated a significant disinfection properties on stagnant water frequently used for irrigation, more than Alum.This observation has been extensively reported for Moringa (Yongabi et al., 2010) but not with the other plant materials used in this study.

Type of Treatment	THBC Cfulml	Coliform counts Cfulml	E Coli counts Cfulml
Untreated water sample initially collected	TNTC	TNTC	8.900
Untreated water Sample left to settle and supernatant analysed	TNTC	TNTC	7.900
Alum treated	3,598	1,380	980
Moringa Oleifera seed treated	485	298	125
Jatropha seeds treated	2,212	598	386
Garcinia Kola treated	387	452	294
Carica papaya seed treated	868	483	223
Persea americana seed treated	1,201	822	429
Hibiscus sabdariffa seed treated	258	205	110

* TNTC = Cful/ml>10.000

Table 6. Effect of plant seed powders and Alum on stagnant water from a dirty pond and used for irrigation of crops at Abubakar Tafawa Balawa University.

In table 7, data indicates significant changes in pH and turbidity when plant materials are applied in a 24 -72 hours retention time.

Treatment materials	Turbidity assessment	PH range	Other Remarks
Retention time 24 - 72 hours			
Untreated water sample initially collected	No floc formed at all	7.0	bad odour peceived
Untreated water sample left to settle and supernatant analysed	Few particles stuck to the wall of the container	7.0	
Alum treated	Flocs formed and settled	5.0	Odour (faint)
Moringa Oleifera treated	Flocs settled and good settlement	7.0	Very clear and second stage treatment very good, odour no trace.
Jatropha seed treated	Flocs formed particles settle	7.0	Odour of Jatropha
Garcinia Kola treated	Flocs settled with suspended pellicle.	7.0	Water clear and a second stage treatment clearer no odour.
Carica papaya seed treated	flocs formed slowly	7.0	papaya odour
Persea americana seed treated	flocs settlement seen slowly	7.0	no dour
Hibiscus sabdarrifa seed treated	flocs settled at the bottom	5.0	Little odour

Table 7. Physicochemical properties of Treated and Untreated stagnant water used for irrigation of crops at Abubakar Tafawa Balewa University.

In table 8, the results show a significant level of disinfection of storm water by the plant materials in comparison with Alum.This findings is in tandem with a similar findings using Hibiscus, Moringa and Jatropha by Yongabi et al., 2011

Types of Treatment	THBC Cfulml	Coliform Cfulml	E Coli Cfulml
Untreated storm water sample initially collected	TNTC	TNTC	TNTC
Untreated storm water sample left to settle and supernatant analysed (24hours)	9.280	212	36
Alum treated storm water	9.200	60	0
Jatropha seeds treated water	6,930	180	20
Moringa Oleifera seeds treated storm water	120	40	11
Garcinia Kola seeds treated water	6,33	160	18
Carica Papaya seeds	398	29	5
Perseas americana seeds	5.360	64	0
Hibiscuss sabdarrifa	4,024	50	32

* 1.   Strom water harvested in flowing through the dirty streets of Yelwa after heavy rains.
   2.   5g of each the seed powder Alum into 100mls of the wastewater and left on bench for 24 hours.

Table 8. Effect of plant seed powders and Alum on Storm water collected from Bauchi Metropolis.

Not only did the microbial content changed after treatment with the natural materials but the pH and turbidity also changed considerably as shown in the data in table 9.

Treatment Materials Retention time 24 - 72 hours	Turbidity assessment	PH range	Other remarks
Untreated storm water sample initially collected	No floc formed no settlement of particles, brownish and dusty smell	6.9 - 7.0 -	-
Untreated storm water sample left on bench for 24 hours	No floc formed, few particles settle at the bottom, supernatant still has suspended particles. Some stuck to the walls of the container.	6.9 - 7.0	-
Alum treated storm water	Floc formed and fast, water clear supernatant clear	5.0	Clean and clear standard coagulant (1st)*
Moringa Oleifera seeds treated storm water	Flocs formed when seeds dispersed in water flocs settled slowly supernatant	7.0	Moringa mildly extracts in water. Good Coagulant (2nd)*
Jatropha Curcas seeds treated	Flocs formed gently and settle	7.0	Good Coagulant (4th)*
Garccini Kola seed	Flocs formed, particles settled, very good coagulant at the bottom Excellent	6.9-7.0	Good Coagulant (3rd)*
Carica papaya seed	Flocs formed, particles settled	6.9-7.0	Good Coagulant (6th)*
Persea americana Hibiscus sabdarrifa	Not very clear excellent, particles Excellent, particles settled flocs settled, good coagulant	6.9-7.0 5.0-5.0	not a good coagulant (5th)* Good Coagulant (5th)*

Table 9. Physicochemical properties of treated and untreated storm water with Alum and plant seed powders.

In the table 10 below, combined plant material with clay was applied in the treatment of refinery wastewater.this hubrid plant and geological material significantly improved the water quality bacteriologically and physicochemically than with the application of just either of the materials alone.

Type of Treatment	Colour	PH	Smell	Appearance	THBC	Coliform	Ecoli	TFC
untreated OVH	colourless	6.6	Engine oil, crude oil smell	Clear	560	Nil	Nil	315
Treatment with Moringa Oleifera seeds	Very clear no odour	7.0	no Odour	clear	36	Nil	Nil	100
Treatment with clay	clear	7.0	no odour	clear	32	nil	nil	96
Untreated DFH	Brownish	7.05	Strong engine oil smell	Turbid	300	nil	nil	6140
Treament with Moringa Oleifera seeds	very clear no odour	7.0	Odour absentr	clear	96	nil	nil	89
Treatment with clay powder	Clear	7.0	Odour absent	clear	93	Nil	Nil	84

Table 10. Effects of Combined Moringa Oleifera seed powder and clay in the treatment of oil refinery wastewater

In the table 11 below, a combined plant material comprising plants (moringa oleifera seed powder) and sand filter media was applied to treat refinery wastewater and the results indicated a significant improvement in water quality both bacteriologically and phyicochemically better than with either of the materials alone. This corroborates a similar observation using surface water in Cameroon (Yongabi et al., 2010 and yongabi et al., 2011)

Type of Treatment	Colour	PH	Smell	Appearance	THBC	Coliform	Ecoli	TFC
untreated OVH	colourles	6.6	Engine oil, crude oil smell	Clear	560	Nil	Nil	315
Treatment with Moringa Oleifera seeds powder	Very Colourless	7.0	Odour absent	very clear	36	Nil	Nil	100
Final treatment with sand filter	Colourless	7.0	Odour absent	Very clear	3	nil	nil	nil
Untreated DFH	Brownish	7.05	strong engine oil smell	Turbid	300	nil	nil	6140
Treatment with Moringa oleifera seed powder	Very clear no colour	7.0	Odour absent	clear	96	nil	nil	89
Final treatment with sand filter	Very clear	7.0	odour absent	clear	10	nil	nil	6

Table 11. Effect of combined Moringa Oleifera seed Powder and sand filter media on oil refinery waste water

The data in table 12 below also shows similar findings using Garcinia kola sand filter media as with moringa oleifera sand filter media.

Type of Treatment	Colour	PH	Smell	Appearance	THBC	Coliform	Ecoli	TFC
untreated OVH	colourles	6.6	Engine oil, crude oil smell	Clear	560	Nil	Nil	315
Treatment with Garcinia Kola seeds powder	Very Colourless	7.0	Odour absent	Clear	70	Nil	Nil	173
Final treatment with sand filter media	Very colourless	7.0	Odour absent	clear	15	nil	nil	8
Untreated DFH	Brownish	7.05	Strong engine oil smell	Turbid	300	nil	nil	6140
Treatment with Garccinia Kola seed powder	Clear very little odour	7.0	no odour	clear	89	nil	nil	125
Final treatment with sand filter media	Clear no colour	7.0	no odour	clear	13	nil	nil	29

Table 12. Effects of Combined Garcinia Kola seed powder and sand filter media on oil refinery wastewater

In tables 13, 14, 15, 16 and 17 the combined performance of the plant materials and clay, and cobined plant materials and sand filtered on various polluted water samples was tested bacteriologically and physicochemically.The results generally indicated strongly that these natural materials have strong ability to purify any type of water.The materials alone have the ability to treat water and wastewater but the combined effect of these materials have an added advantage in treating all kinds of polluted water as demonstrated by the data in the following tables 13, 14, 15, 16 and 17.

Type of Treatment	Colour	PH	Smell	Appearance	THBC	Coliform	Ecoli	TFC
untreated waste water sample (detergent)	Bluish dirty	6.5	acrid smell	Turbid and foamy	2,200	2,300	1,900	-
Treatments with Moringa seed powder	Blue colour fades away	7.0	odour absent	flocs formed	320	520	343	-
Treatment with clay	Clear	7.0	odour absent	clear needs filtration	309	511	338	-

Table 13. Effects of Moringa Oleifera seed powder and clay in the treatment of detergent based waste

Type of Treatment	Colour	PH	Smell	Appearance	THBC	Coliform	Ecoli	TFC
untreated OVH	colourless	6.6	Engine oil, crude oil smell	Clear	560	Nil	Nil	315
Treatment with Hibiscus Sabdariffa seed powder	Colourless	5.0	odour faint	clear	133	nil	nil	113
Treatment with sand filter media	very colourless	5.0	total odour removal	very clear	5	nil	nil	15
Untreated DFH	Brownish	7.05	Strong engine oil smell	Turbid	300	nil	nil	6140
Treatment with Hibiscus Sabdariffa seeds	clear	5.0	little odour	clear	118	nil	nil	595
Treatment with sand filter media	clear	5.0	No odour	clear	3	nil	nil	20

Table 14. Effects of Combined Hibiscus sabdariffa seed powder and sand filter media on oil refinery wastewater

Type of Treatment	Colour	PH	Smell	Appearance	THBC	Coliform	Ecoli	TFC
untreated OVH	colourless	6.6	Engine oil, crude oil smell	Clear	560	Nil	Nil	315
Treatment with carica papaya seeds	Colourless	7.0	odour absent (papaya scent)	clear	62	nil	nil	87
Treatment with sand filter media	very colourless	7.0	odour absent	very clear	2	nil	nil	5
Untreated DFH	Brownish	7.05	Strong engine oil smell	Turbid	300	nil	nil	6140
Treatment with Carica papaya seeds	Clear very little odour	7.0	Little odour a bit of papaya scent	clear	90	nil	nil	95
Treatment with sand filter media	Clear	7.0	no odour	clear	nil	nil	nil	10

Table 15. Effects of Combined Carica papaya seed powder and sand filter media on oil refinery

Type of Treatment	Colour	PH	Smell	Appearance	THBC	Coliform	Ecoli	TFC
untreated waste water sample (detergent)	bluish dirty	6.5	acrid smell	Turbid and foamy	2,200	2,300	1,900	-
Treatment with Moringa Oleifera powder seed	flocs formed colour removed	7.0	odour removed totally	clear, no foams	320	1'070	343	-
Treatment with sand filter media	clear	7.0	no odour	clear	-	-	-	-

Table 16. Effects of Combined Moringa Oleifera seed powder and sand filter media on detergent based

Type of Treatment	Colour	PH	Smell	Appearance	THBC	Coliform	Ecoli	TFC
untreated wastewater sample (detergent)	very foamy bluish	6.5	acrid smell	Turbid foamy	2,200	2,300	1,900	-
Treatment with Garcinia Kola seed powder	flocs formed with a suspended pellicle	7.0	smell reduced	becoming clear	700	675	521	-
Final treatment with sand filter media	clear	7.0	odour absent	clear	100	5	1	-

Table 17. Effects of Combined Garcinia Kola seed powder and sand filter media on detergent based water

In table 18 below, one of the plants:Moringa oleifera was used to study its effect on unicellar organisms in water.The results indicated that moringa oleifera seed powder gets rid of unicellular organisms such as amoeba, microalgae such as spirogyra from water.The need to study the application of plant materials in the removal of microalage from water systems could be rewarding.

Types of Organisms	approximate number per field
Diatoms	up to 15 per field
Cercaria	a few
Euglena	35 cells per field, actively motiles
Cyclops	a few
Amoeba	More than 15 per field
Debris	a lot of debris
Spirogyra	a lot

a) Microscopy of pond/stagnant water before treatment

Types of Organism	approximate number per field
Euglena	totally absent, water clean
Diatoms	absent
spirogyra (blue/green algae)	absent
Cyclops and cercaria	absent
Amoeba	absent

b) Microscopy of Pond/Stagnant water after treatment with Moringa Oleifera seed powder, and after filtration

Table 18. Effects of Moringa Oleifera seed powder on free living organisms in pond water used for irrigation

An novel attempt was made to classify materials that can be applied in water pollution management and shown in table 19 below.more studies for a detail classification are underway

Botanical name of plants	Common name/ Hausa name	Part used	Types of wastewater	Types of Coagulant	Sources
Jatrohopa Curcas	Physis not Benin Zugo	Seeds	Industrial effluents domestic wastewater	Phyto Coagulant	Yongabi, K.A (2004)
Sychonos Potatotum	-	Seeds	Domestic water	Phyto-Coagulant	-----
Moringa Oleifera	Horse- raddish Zogale	Seeds	Domestic water Industrial wastewater	Phytocoagulant "	Pers.Comm. Yongabi, K. A
Calotropis procera	tumfafiya	latex	Wastewater	"	Pers. comm.
Citrus aurantifolia	Limes, lemu	seeds	domestic water	"	-------
Pumice	Rock	-	domestic and industrial wastewater	Geocoagulant	Internet
Bentonite	Rock	-	"	"	"
Immansil	"	-	"	"	"

Table 19. Survey and classification of Natural materials for water pollution management in local communities

In table 20, the nature of extracts from Garcinia kola was described. The water extract is a black solid.The coagulant and disinfection activity of Garcinia kola observed in this study may be soluble in water.More studies are need in this dimension.

Solvent	Boiling point	Volume of Sovent (ml)	Nature of Extract	Crude yield (g)
Dichloromethane	-	-	-	-
N-Hexane	69°C	250	Golden yellow oily semi solid	0.90
Toluene	III°C	250	Dark yellow oily semi solid	2.10
Acetone	56°C	250	Dark semi solid	1.90
Methanol	65°C	250	Dark brown solid	1.70
Water	100°C	250	Black solid	1.10

Table 20. Analysis of Phytochemical tests on solvent extracts of Garcinia Kola

The results in table 19 gives an attempt to classify some of the phytoconstituents in this plant materials.More phytonutrients were detected in the aqueous extract suggesting an easy and cheap means of extracting water treatment chemicals from Garcinia kola.

Solvent Extract	Cardiac Glycosides	Saponin	$C_6H_{12}O_6$	Tannins	Flavoniods	Alkaloids
Dichloromethane	-	-	-	-	-	-
n-Hexane	+	-	-	-	-	-
Toluene	-	-	-	-	-	-
Acetone	-	-	-	-	-	-
Methanol	-	-	+	-	+	-
Water	+	+	+	+	-	-

Table 21. Result of Preliminary phytochemical analysis of Solvent Extracts of Garcinia Kola

The data in table 22 shows that plant materials can significantly stabilize pH of various polluted water.This has ben observed with moringa in previous syudies (Yongabi et al., 2010) but has not been done using various wastewater samples such as from cement and asbestos.

Types of water /waste water	PH (Normal)	PH Alum (treated)	PH Moringa treated	PH Garcinia (treated)	PH Hibiscus treated	PH Carica treated	PH Jatropha
Dirty tap water	6.62	5.0	7.0	6.99	5.0	7.0	7.0
Yelwa tap Water	7.36	5	5.0	7.0	7.0		
University tap water	7.25	5.0	7.0	7.0	5.0	7.0	7.0
Yelwa well water	7.37	5.0	7.0	7.0	5.0	7.0	7.0
Asbestos water (well)	7.46	5.0	7.0	7.0	5.0	7.0	7.0
Asbestos tap water	7.53	5.0	7.0	7.0	5.0	7.0	7.0
Cement waste water	8.01	5.0	7.0	7.0	5.0	7.0	7.0
Cement waste water	8.52	5.0	7.0	7.0	5.0	7.0	7.0
Cement waste water	8.50	5.0	7.0	7.0	5.0	7.0	7.0

Table 22. PH Content of various waste water/water treated with Alum and plant seed powders

To further demonstrate the disinfection potential of the plant materials on the wastewater samples, a methanol extract of the plant materials was conceivable.The resulting extracts were tested on various bacterial isolates from all the polluted water samples.The data in table 21 below demonstrates a significant level of antibacterial activity comparable to Alum

Extracts	E coli	Pseudomonas Sp	Klebsiella Sp	Staphyloccus
Garcinia Kola seeds Aquenus Extract Methanol Extract	60mm	6mm	15mm	18mm
Hibiscus sabdariffa seeds Aquenue Extracts	10.8mm	12.0mm	12mm	15mm
Methanol Extracts	11mm	11.8mm	8mm	19mm
Carica papaya Seeds Aqueous Extract	9mm	12mm	14mm	16mm
Methanol Extract	11mm	12.5mm	16mm	20mm
Aluminium Sulphate	12mm	13mm	10mm	10.5mm
Water	0mm	0mm	0mm	0mm
Methanol	5mm	9mm	11mm	8mm

Table 23. Effect of cold Methanol and Aqueous Extract of Garcinia Kola, Carica papaya and Hibiscus Sabdarrifa seeds on Bacterial isolates from waste water (Diameter zone of inhibition in mm)

CFU	Colony Forming Units
COD	Chemical Oxygen Demand
CWE	Crude Water Extract
IFX	Ion Exchange
MIC	Minimum Inhibitory Concentration
MO	Moringa Oleifera
MOCP	Moringa Oleifera Coagulant Protein
UC	Uniformity Coefficient
OD	Optical Density
WHO	World Health Organization
WPC	Water Production per cycle
TNTC	Too numurous to count.
Ml	Mililitre
OVH	Overhead Fraction or Foul Water
DFH	Desalter Foul Water

5. A pilot water treatment plant using natural materials at government technical College, Njinikom, Bamenda, Cameroon

The Phytobiotechnology Research Foundation (PRF), Cameroon, in collaboration with the School of Chemical Engineering, The University of Adelaide, South Australia, is proposing to carry out a capacity building training on:A simple Moringa- sand based water filtration technology for clean potable water supply in the rural schools and villages in Boyo Division, Cameroon.This is part of a doctoral research in chemical engineering, The University of Adelaide, south Australia.Three undergraduate students in chemical engineering are undertaking their honours thesis on the water quality , management and training, safety and ethical issues associated with the implementation of Integrated biocoagulant-sand filter system for drinking water purification at Government technical college,

Njinikom, Cameroon. A well with an approximate water volume of 2500 litres has been dug, and a filtration system using Moringa oleifera seeds and sand filter is being constructed expected to purify 2000 litres of water in 24 hours retention time to serve more than 7000 students.

5.1 Anticipated benefits
- Clean potable water will be available for rural people.
- Decimation of incidence of infectious/waterborne diseases

Improved health

5.2 Training method
The training shall be conducted in conjunction with local NGOs in Bamenda, Cameroon. PRF is an NGO based in Bamenda and has a track record on community development projects in Cameroon and Nigeria. PRF has entry points to communities and has over the years worked with a number of Research institutes in the country. PRF has facilitated a number of training for local groups in Bamenda water quality in rural areas.Similarly, PRF has participated at training on water filtration technology at the ZERI Centre in Nigeria.

Five (5) selected people from the local schools shall be trained and then the school authority shall provide them the resources to mount the outfits .The students will be encouraged to set up a household filter unit in their homes during holidays.

These trainees shall function in union with the PRF and the school authority who will in turn monitor and supervise effective functioning of the filter units.

6. Conclusion and recommendation

The research work has shown that there are many natural materials available in many communities in the world that can be used to treat water for drinking.Additionally, this research has demonstrated that these plant and geological materials can be applied in the treatment of any type of polluted water.These materials are ecological , low cost when compared to the application of synthetic chemicals currently used in water pollution management.The ongoing pilot system applying natural materials in water treatment in Cameroon could be replicated elsewhere.More research into the use of natural materials in water pollution management should be studied.

7. Acknowledgement

The authors would like to thank the University of Adelaide, South Australia for a phD scholarship to do this work.The Phytobiotchnology research Foundation , Cameroon and the Principal of GTC njinikom, Cameroon for provision of funds to set up the pilot work and research.

8. References

Eilert, U, Wolters, B. and Mahrstedt, A. (1980) Antibiotic Principles of seeds of Moringa Oleifera' Planta med., 39, 235.

Elest, U., Wolters, and Mahrstedt, A. (1981). The Antibacterials Principles of seeds of Moringa Oleifera and Moringa Stenopetala: Planta Medica, 42 (1), 55)

Gopala Krishna, K. S. Kurup, P.A., and Narasimba - Rao, P.l. (1954) Antibiotic Principles from Moringa Pterygosperma. Part III Action of pterygospermin on germination of seed and filamentous fungi; Indian J. med. Res, 42, 97-99

Kurup, P.A., and Narashima - Rao, P.L. (1954a) Antibiotic Principles from Moringa pterygosperma. Part V Effect of pterygospermin on the assimilation of glutamic acid by Micrococcus pyogene var. dureus, Indian J. Med. Res., 42, 109-13

G.K. Folkard, J.P. Sutherland, M. A., Mtawali and W.D., Grant Moringa Oleifera as a Natural Coagulant, http // info.lut. ac.uk/departments/cv/wedc/garnet/wares.html University of Leicester, UK.

Olson, M.E and S.G. Razafimandimbison (in Press) M hildebrantii; a tree extinct in the wild preserved By Indian horticultural practices. M. Peregrina (www.fao.org/decrop/ r775oe/r7750eo4.htm)

www.le.uk/engineering/staff/sutherland/Moringa/cultivation/cult/htm

Oyawoye, O. M., Ogbadu, L. J., Abayeh L. J. Abanyeh, O.J. and Agbo, E.B (2000) Distribution of Nitrate in Drinking waters of Bauchi in press.

Adegbola (1987) The impact of Urbanization and industrialization on health conditions.The case of Nigeria.World health statistics Quarterly, 87 (40):74-83

APHA (1950 Water supply- nitrate in potables waters and Methemoglobinemia 'Year book of the American Public Health Association (APHA) New York.

American Public Health Association (APHA) (1988), Standard Method for the Examination of Water and wastewater. Many Ann Franson editions. 15th Ed. Washington DC. Repress Spring Field.

Dada, O.O. Okuofu, C.A. and Yusuf, T.R. (1990) the relationship between chlorine residual and Bacteriological quality of tap water in the water distribution system of Zaria, Nigeria, Savana 2(1), 95-101

Dzwairo, B., Hoko, Z., Love, D and Guzha, E (2006) Assessment of the impacts of pit latrines on ground water quality in rural areas: a case study from marondera district, Zimbabwe. Physics and Fuchs.

Bina (1991) Investigation into the use of Natural plant coagulants in the removal of bacteria and bacteriophage from turbid waters, PhD thesis, University of New Castle Upon Tyne

Who (1985) Guidelines for drinking water quality vol3: Drinking water quality control in Small Community Supplies. Who, Geneva, P. 47-121.

Sameer, F.J., and Ameh, K.A. (1986) Bacterial Contamination of Drinking water Supplies in Baghdad City, Iraq, JBSR 17 (2): 313-315.

Sandhu Shigara S: William J. Waren, and Peter Nelson (1979) Magnitude of pollution indicator organisms in rural potable water. J. Appl. Environ Microbiol 37(4): 744-749.

White, G.C. (1972) Handbook of Chlorination, New York: Van Nostrad Reinhod P. 60-82

Houssain Abouzaid, (1988) Evaluation of Coliphage and Presence/Absence tests for the sanitary Classification of the water Resources and the quality of Drinking water in Morocco. Office National de l'eau potable (ONEP) B.P. Rabat-Chellah.

Cheesbrough, N (1984) Medical Laboratory Manual for Tropical Countries, Tropical Health Technology, Butterworth, pp 1-15.

United Nations Food and Agricultural Organization (FAO) 1996 for all, Rome, FAO: P.64

Paul, G (1998) Uprising: The Road to Zero Emissions. England: Greenleaf publishing.

Jackson, A.R.W. and Jackson, J.N. (1996) Environmental science; the natural environment and Human impact. Singapore: Longman Group Limited.

UNICEF (1993) Control of diarrhea diseases (CDD) adapted from facts of life, Watsan Health Qi.W; He, Y. Wei, F. and fang, X. (1983). Nickel Contamination in the homes of employees of Secondary nickel smelters. Environmental Research 15:373-380.

UNICEF (2009) Soap, Toilets and taps, A Foundation for healthy Children, How UNICEF supports water, sanitation, hygiene:http//wwwunicef.org/wash/files FINAL-showcase-doc-for web.pdf

UNEP (2002) Past, Present ad Future perspectives, Africa environment outlook.United Nations Environment Programme, Nairobi, Kenya.

Aziz-Alraham, A.m; Al -hajjaji, A.M and Al Zamil (1984). Environmental impact of heavy metals. Journal of Environmental Health. 40; 306-310

Adam, J (1983). The effects of air pollution on plants and animals. Microchemical Journal of science 28(1):82-86

Tinslay, D.A., Baron, A.R., Critchley.R and will-Lamson, R.J (1984). The fate of heavy metals in: Greenland, DJ and Hayes, MHB (eds). The Chemistry of soil processes. Chichester, John Wiley and Sons, pp 593-620.

Crecelius, E.A., Johnson, C.J. and Hofer, G.C. (1974) Contamination of soils near a copper Smelter by arsenic, centimony and lead. Water, Air, Soil pollution. 3:337-342.

Shack lette, H.J. (1972). Distribution of trace elements in the environment, and the occurrence of heavy disease in Georgia U.S. geological society of America. Special paper. 140:65-70.

Madore, M.S, Rose, J.B., Gerba, C.P, Arrowood, M.J and Sterling, C.R (1987) Occurrence of Cryptosporidium oocysts in sewage effluents and selected surface waters.Journal of Parasitology, 73:702-705

Morton, W.E. and Dunette, D.A (1994) Health effects of Environmental arsenic. In; arsenic in the Environment, Part II; Human health and Ecosystems effects, wiley and sons, New York pp 17-34.

Smith, A.H. Hope n-rich, C., Bates, M.N. Gorden, H.m., Hertz-Accioti, I., Duggan, H.M.wood, R., Kaswett, M.J. and Smith M.T. (1992). Cancer risks from arsenic in drinking water. Environ. Helath perspectives 97, 259-269.

Schulzt , C.R and Okun, D.A (Surface water treatment for Communities in developing Countries. Journal of American Water Works Association, 75:212-219

Tseng, W.P. (1989). Black foot disease in Taiwan; a 30 year follow-up study. Angio 40, 547-558.

Etherton. A.R.B (1975), (1975), Mastering Modern English, A Certificate Course, Hong Kong: Longman 178-179.

Pritchard, M; Mkandawire, T;Edmondson, A;O'Neill, J.G and Kululanga, G (2009) potential of using plant Extracts for purification of Shallow well water in Malawi.physics and Chemistry of the Earth, 34: 799-805

WHO (2006) Guidelines for drinking water quality.First Addendum to the third edition, recommendations, Vol.1.http//www.who.int/water_sanitation_health/dwq/gdw0506.pdf

Yongabi Kenneth, Lewis David and Harris Paul (2010) Alternative Perspectives in Water and Wastewater treatment.Lambert Academic Publishing, PP 127, ISSBN 978-3-8383-8785-7

Yongabi, K.A;Lewis, D.M and Harris, P.L (2011) Application of Phytodisinfectants in Water treatment in rural Cameroon.African Journal of Microbiology Research, Vol.5 (6) pp 628-635

9

The Relationship Between Metal Forms Found in River Bottom Sediments and Land Development (Review)

Anna Rabajczyk

The Jan Kochanowski University of Humanities and Sciences in Kielce
Independent Department of Environmental Protection and Modelling, Kielce
Poland

1. Introduction

Components of the environment comprise three-phase aquatic ecosystems, whose basic phase is the liquid, i.e. water. On the one hand, it borders the gas phase, which is the atmospheric air, on the other – the solid phase, i.e. the ground which is most frequently covered with bottom sediment. In addition, water contains suspensions which may be formed by inorganic matter and living organisms, representing both flora and fauna. Consequently, the sources of origin and migration routes of various substances which affect surface water quality are numerous.

An integral part of the aquatic environment is bottom sediments. They play a significant role in the biochemical cycle of elements as the place of deposition and chemical transitions of many compounds which find their way into water. Moreover, they constitute a habitat of water organisms, which results in biochemical transitions of the compounds deposited in the sediments. An assessment of pollution impact on surface water quality requires a study of anthropogene concentration levels and sources. In this respect, bottom sediments which function as sorption column offer particularly useful material for research aimed at determination of major sources of pollution as well as providing a clear picture of what happens in the pelagic zone. Given the relatively easy metal migration from the environment to the water phase, it may be assumed that the metal concentration level in bottom sediments is a sensitive indicator of environmental cleanliness.

Together with water, soluble and insoluble substances in various states of matter migrate and may encounter resistance from a large variety of factors. The latter are usually taken to include internal and external friction, changes in shapes and cross-section dimensions, river load separation and transport, local obstacles to water flow in the river channel, vegetation overgrowing the river channel, as well as irregularities and curvatures of the horizontal system. The total resistance of the river channel is the sum of partial resistances due to individual factors, but the proportion of a factor in total resistances varies depending on flow volume as well as season, which need to be considered in hydrological analyses.

For research into the processes which occur in surface waters, particularly those which shape bottom sediments, significant is the issue of direct measurement of river load movement intensity. Accurate understanding of the processes which occur in the river load, including erosion, and transport – together with the accumulation of its material – is especially important, chiefly for determination of the speed and route of the pollution accumulated in the sediment [1].

Metals bonded through physical adsorption and chemisorption are in the state of equilibrium with the intermolecular water of bottom sediment and can very easily penetrate into the solution which occurs mainly at increasing water salinification (in particular, an increase in chloride ion content, the ions complexing some of the metals). Metals, in turn, which are bonded through co-precipitation with hydrated iron and manganese oxides and carbonates penetrate into the solution with greater difficulty. This can occur under strongly reductive conditions in the case of metal oxides, and in the case of carbonates – under considerable environmental acidification [2].

Primary and secondary (i.e. of pollution origin) metals are found in environmental material in various chemical forms. In view of that, an assessment of their mobility ought to include several aspects. The content of mobile metal forms which can take direct part in the biocycle, including soluble, ion-exchangeable and unstable complex metal associations, should be determined. An equally important issue is the content of solid metal forms as a potential reserve in the biocycle (e.g. chemisorption ions, poorly soluble salts or metal forms which are part of complex associations with organic matter) and isomorphic contaminants of various metal forms, the siliceous in particular [3].

Considering the relatively easy migration of metals from the environment to the water phase, it may be assumed that the metal concentration level in bottom sediments is a sensitive indicator of environmental cleanliness. Yet in order to obtain information on potential metal mobility in the investigated environment as well as assessing environmental impact of pollutants contained in the sediment which may be removed therefrom as a result of changing physical, chemical or biological conditions [4], it is necessary to perform an operational speciation analysis (fractionation).

A speciation analysis of compounds in aquatic ecosystems, performed with a focus on bioavailability of individual entities and forms in which they occur, provides an opportunity to obtain the data necessary to define environmental hazards. The forms in which elements occur condition their toxicity, synergism and antagonism, or lack of these in relation to other elements, and hence – positive or negative impact on the functioning of living organisms. In a discussion of pollution-related problems, it is important to differentiate between various oxidation states and forms in which metals occur, both at the stage of their penetration into the environment, their migration and changes to particular elements of the biosphere.

Studies of bottom sediments, both in watercourses and in reservoirs, which have been conducted for several years, point to various forms of metals cumulated at the bottoms of aquatic ecosystems. Depending on biological, chemical and physical properties of the environment, elements such as Cd, Cr, Cu, Fe, Pb, Ni and Zn can occur in a variety of forms and be related to, for example, organic or oxide matter. Some studies show, that oxic sediments tend to accumulate transition metals and a few other elements, e.g. Mn, Cu, ni, Co, V and Mo, while anoxic sediments overlain by bottom waters with very low or no oxygen accumulate these metals, with exception of Mn, plus many of the otherwise relatively unreactive oxyanions [5-7].

The interest in the sequential extraction technique for studies of metals in water sediments has been consistently growing, and more than ten methods are in use presently, enabling the separation of individual element fractions. The methods include those of Gatehouse et al. [8], Tessier et al. [9], Sposito et al. [10], Ure et al. [11] and Hall et al. [12]. To characterise functional speciation of metals in bottom sediment, Tessier's five-stage method [9] which enables the separation of five fractions is most commonly applied. Consequently, most frequently analysed are the fractions in which metals occur in the following forms:

- exchangeable (F1): these are metals adsorbed on the surface of solids which can pass to the solution due to changes to ionic composition of water or shift of balance in the sorption-desorption system; this fraction is available and relatively mobile;
- bonded to carbonates (F2): these are metals which occur in the form of carbonates or are co-precipitated with carbonates; as a result of decreasing pH values, the carbonate balance is disturbed which causes passage of metals to the solution;
- bonded to hydrated iron and manganese oxides (F3): these are metals adsorbed on the extended surface of the precipitating hydrated iron and manganese oxides under anaerobic (reductive) conditions; due to iron and manganese reduction, the sediment can be dissolved and metals can pass into the solution;
- bonded to organic matter (F4): these are metals adsorbed on the surface of organic matter or metals bonded to that matter; they are temporarily immobilised, but due to naturally progressing sediment mineralisation, they can – with time – pass to one of the other fractions;
- permanently bonded with minerals (F5): these are metals built into the crystal network of both secondary and primary minerals; they are permanently immobilised and, under natural conditions, do not pose a threat to the ecosystem [9].

The physical and chemical properties of aquatic ecosystems are characterised by several mutually dependent parameters. Depending on factors such as geological structure, land development, the form in which individual pollutants migrate to the aquatic ecosystem, climate factors (including temperature and humidity), the amount of oxygen or the environment's acidity, change may occur in solubility of salts present in the waters, forms in which individual entities occur, their bioavailability and toxicity. Consequently, familiarity with the mechanisms of heavy metal mobility in the aquatic environment, the specification of conditions which may cause re-mobilisation of substances deposited in bottom sediments into the environment and the ensuing assessment of real time-delayed chemical hazards is a very significant issue for aquatic ecosystem quality maintenance as well as elimination of aquatic ecosystem pollution sources.

It should be noted that few of heavy metals, like manganese and iron, are of particular importance to river system, moreover manganese can be the most important oxidant in environmental. Fe and Mn cycling is a key thermodynamic regulator and kinetic catalyst in natural waters and their presence is connected with the oxygen, nitrogen, carbon, sulfur, and phosphorous geochemical cycles [13-15].

The transport and fate of heavy metal in natural waters is strongly affected by Fe and Mn oxide precipitation and dissolution. Heavy metals adsorb on Fe and Mn oxide, create and determine a form of physical and chemical properties of the sediments. The heavy metals are also incorporated in the Fe and Mn oxide matrix as impurities when precipitation occurs or when new mixed metal/Fe and metal/Mn coprecipitates are possible [15, 16]. Aqueous Fe(II) and Mn(II) are significant in natural waters only in the absence of O_2. Insoluble Fe(III) and Mn(III/IV) oxides form under oxic conditions. But manganese-oxide reductive dissolution is generally thought to occur concurrently with the reductive dissolution of Fe oxides [17]. Their solubilities limit the aqueous concentrations of Fe and Mn species. The metamorphosis among redox states and physical states are frequently slow in the absence of catalysis, e.g. aqueous solutions of Mn(II) in the presence of O_2 at pH=8.4 are exoergic toward oxidation, yet the uncatalyzed reaction proceeds slowly across years [15].

Long-term deposition of compounds such as iron and manganese oxides and carbonates in the sediment may lead to these substances being turned into permanent crystal structures, which can cause total exclusion of the co-precipitated heavy metals from the cycle. Another situation occurs when metals are bonded to an organic substance. Its decomposition may cause passage of metals into the solution or their transition into other insoluble forms. Considerable amounts of metals may be trapped in the silicate crystal network, or form poorly soluble compounds. In practice, these metals cannot pose a hazard to the environment under natural conditions [18].

2. Forms of metals in sediments of selected rivers

From the chemical viewpoint, metals in sediments can occur in the forms of carbonates, hydroxides, silicates, sulphides, phosphates, or compounds with organic ligands at various crystallisation stages of various stechiometry and water content [19]. Moreover, some of these associations may be adsorbed on larger molecules. However, a change in salinity or increasing concentration of the metal-complexing Cl⁻ is sufficient for the adsorbed substances to pass into the pelagic zone. The metals bonded to organic matter, upon its decomposition, are also released while those found in silicate crystal networks do not actually pass to the pelagic zone [18].

Climate, development of the catchment area, the geological substratum, and the river type are but some of the factors which have affected river quality. Research into metal speciation in bottom sediments conducted in various parts of the globe suggests that each aquatic ecosystem ought to receive individual treatment. Even various bottom sediment sampling sites within a single river are affected by different factors (Table 1).

A case in point may be the studies into bottom sediments of the Bobrza River, a right tributary of the Czarna Nida River, conducted by the present author [28, 29]. The river flows across terrain built of middle-Devonian limestone and dolomite, and across a depression filled with Pleistocene sediments of fluvial and glacial accumulation. This terrain is characterised by a large density of industrial plants and mines (cement and lime industry); closeness to a railway line and communications routes which are a source of dust pollution emissions, containing both metallic and non-metallic elements; as well as numerous tributaries which receive municipal and industrial wastewater. Sampling sites were located in front of and past the area where cement and lime industrial plants are situated.

The investigated river contains no natural, related to substratum structure, sources of pollution with zinc, cadmium or iron. Consequently, the presence of metals in the aquatic ecosystem is a result of human activity. The total metal content in the bottom sediment is evidence of the bottom sediments being burdened with Zn, Cd and Pb from anthropogenic sources. The Bobrza River bottom sediments are characterised by large amounts of Zn, Cd and Pb in the forms of carbonates or precipitating hydrated iron and manganese oxides, co-precipitated with carbonates and adsorbed on the extended surface. Considerably lower analyte amounts are found in Fractions F1 (exchangeable) and F4 (adsorbed on the surface of organic matter or bonded to that matter). The lowest amounts of Zn and Pb are found in the fraction permanently bonded to minerals (F5), and of Cd – in the exchangeable fraction (F1).

River/ Country	Description	Metal	Forms	Ref
Daugava River/ Latvia	The key characteristics of the hydrological regime are high and permanent flood in spring, a summer-autumn high-water caused by rains, lowered levels in midsummer, and low-water periods in winter; composed of Devonian sediments: dolomite, clay, dolomite marlstone, marl, and limestone; land use: about 45% covered with forests, 48% is agricultural land, inland waters- 3%, wetlands –3%, urban areas up 1%	Pb	F5>F4>F2>F3>F1 (FP) F4>F3>F5>F2>F1 (LP)	[20]
		Cd	F5>F4>F3>F2>F1 (FP) F3>F4>F2>F1>F5 (LP)	
Anyang River/ Korea	Limestone and limesilicates also occur along the western side of the river; The pollution of the river water is probably due to the rapid urbanization and industrialization from the early 1970s, giving rise to a large increase in population. Among the industries, the electronic sector comprises the largest part (44%), followed by machinery (24.5%) and chemicals/ textiles (14%)	Pb	F3>F4,F5>F2>F1 (FP) F5>F1>F2>F4>F3 (LP)	[21]
		Fe	F3>F5>F4>F2>F1 (FP) F3>F5>F4>F1>F2 (LP)	
		Cu	F1>F3>F4>F2>F5 (FP) F4>F3>F5>F1>F2 (LP)	
		Cd	F5>F4>F3>F2>F1 (FP) F3>F4>F5>F1,F2 (LP)	
		Zn	F3>F5>F2>F4>F1 (FP) F1>F3>F2>F5>F4 (LP)	
Yamuna River / India	On the basis of the different geological and ecological characteristics, the river has been divided into five segments - Himalayan, Upper, Delhi, Eutrophicated and Diluted - of which Delhi is the most polluted stretch of the river; The river receives treated and untreated effluents from various towns and cities located on its banks. The major industrial towns at its banks before their confluence at Allahabad are Yamunanagar, Delhi, Mathura and Agra	Cu	F1>F3>F2>F4>F5 (FP) F4>F3>F1>F2>F5 (LP)	[22]
		Pb	F1>F5>F3>F2,F4 (FP) F5>F1>F3,F4>F2 (LP)	
		Cd	F4>F5>F3,F1>F2 (FP) F1>F4,F5>F2>F3 (LP)	
		Zn	F1>F3>F2>F4>F5 (FP) F1>F3,F4>F5>F2 (LP)	
Nile River /Egipt	the majority of heavy industry (sugar factories in Komombo, Ques, Armant, Deshna and El-Hawamdia and the oil and Coca-Cola factories in Souha) is concentrated in Greater Cairo and Alexandria; changes in river water quality are primarily due to a combination of land and water use as well as water management interventions such as: (a) different hydrodynamic regimes regulated by the Nile barrages, (b) agricultural return flows, and (c) domestic and industrial waste discharges including oil and wastes from passenger and river boats. These changes are more pronounced as the river flows through the densely populated urban and industrial	Cd	F5>F4>F2>F3>F1 (FP) F5>F2>F4>F3>F1 (LP)	[23-24]
		Cu	F5>F4>F2>F3>F1 (FP) F5>F4>F2>F1,F3 (LP)	
		Cr	F5>F4>F1,F2,F3 (FP) F5>F4>F1,F2,F3 (LP)	
		Fe	F5>F3,F4>F1,F2 (FP) F5>F3,F4>F1,F2 (LP)	
		Pb	F4>F3>F1,F2,F5 (FP) F4>F1>F3>F5>F2 (LP)	
		Mn	F5>F3>F2>F4,F1 (FP) F5>F1,F3>F2>F4 (LP)	
		Ni	F5>F4>F3>F1>F2 (FP) F5>F4>F1>F3>F2 (LP)	
		Zn	F5>F4>F2,F3>F1 (FP) F5>F4>F3>F1,F2 (LP)	

	centres of Cairo and the Delta region			
Ganges River/ India	The river flows in the great alluvial plain, which is of Pleistocene-Holocene origin, and redistributes the weathered sediments of the Gangetic alluvial plain derived from the Himalayas; the river flowing through the districts of Pilibhit, Sajahanpur, Jaunpur and Ghazipur in Uttar Pradesh; there are a few small tributaries; The Gomti river drains over 10 districts. The sediments of the Gomti river are characterized by fine sand with slight changes along the river courses	Cr	F2>F5>F4 (F1,F3-non) (FP) F5>F2 (F1,F3,F4-non) (LP)	[25]
		Cd	F2>F3>F4 (F1,F5-non) (FP) F5>F1 (F2,F3,F4-non) (LP)	
		Fe	F5>F4>F2>F3 (F1-non) (FP) F5>F4>F3 (F1,F2-non) (LP)	
		Cu	F4>F2>F5>F3 (F1-non) (FP) F4>F5 (F1,F2,F3-non) (LP)	
Gediz (G) and Buyuk Menderes (BM)/ Turkey	Rivers are known to be under contamination menace by wastes derived from industrial sources (industrial operations represent approximately 1/5 of the total industrial activity in Turkey), sewage (there are four big cities discharges) and agricultural activities which corespond in these regions to 35% of the total in Turkey	Co	F3>F4,F5>F1,F2 (BM) F3>F4>F5>F1,F2 (G)	[26]
		Cr	F3>F5>F4>F1,F2(BM) F5>F3>F2>F1>F4 F2 (G)	
		Cu	F5>F4>F3>F2>F1(BM) F5>F4>F3>F2>F1 F2 (G)	
		Fe	F5>F3>F4>F1,F2(BM) F5>F3>F4>F2>F1 F2 (G)	
		Mn	F3>F5>F2>F4>F1(BM) F3>F5>F1>F2>F4 F2 (G)	
		Ni	F5>F3>F4>F2>F1(BM) F5>F4>F3>F2>F1 F2 (G)	
		Pb	F5>F3>F4>F2>F1(BM) F3,F5>F2>F4>F1 F2 (G)	
		Zn	F3>F4>F5>F2>F1(BM) F4>F3>F2>F5>F1 F2 (G)	
Asa River/ Nigeria	The Asa River is used for various purposes; domestic, industrial, farming, swimming and so on, both within and outside Ilorin town; there are some industries and establishments located along the course of the river in Ilorin that empty their waste discharges into this river either treated or otherwise. There is conspicuously a soap and detergent Industry, two beverage industries, a major hospital, a major market and a lot of farm practices are carried out along the bank of the river, to mention a few.	Mn	F5>F4>F3>F1,F2 (FP) F4,F5>F3>F1,F2 (LP)	[27]
		Fe	F5>F4>F3>F1,F2 (FP) F5>F4>F3>F2>F1 (LP)	
		Pb	F5>F3>F2>F4>F1 (FP) F5>F3>F4>F2>F1 (LP)	
		Cr	F5>F4>F3>F1>F2 (FP) F5>F4>F3>F1>F2 (LP)	
		Zn	F4>F5>F3>F1,F2 (FP) F4>F5>F3>F2>F1 (LP)	
		Cu	F5>F4>F2>F1>F3 (FP) F4>F5>F1>F2>F3 (LP)	

FP – first point, LP – last point.

Table 1. The metal forms in bottom sediments in selected rivers of the world.

Another distribution of metals in particular fractions has been noted in the studies into the Gorzyczanka River's sediments [29]. The channel of this river is built mostly of mudstone, siltstone, sandstone, shale, quartz and greywacke. Hence in the case of this river again there occur no natural, related to substratum structure, sources of pollution with zinc, cadmium, lead or other metals, with the exception of iron. The channel of the

Gorzyczanka is situated among farming fields, arable land and grassland, with the preponderance of orchards. Due to the crop type, appropriate pesticides and fertilisers are used, the latter containing pesticides, aromatic hydrocarbons, heavy metal salts (including zinc), acids, alkali and phenols. This land development causes metal distribution in individual fractions of bottom sediments to differ from that of the Bobrza River. The highest amount of metals in the Gorzyczanka River sediments has been noted in the organic (Zn) and oxide fractions (Cd and Pb).

Yet another fractionation result has been obtained in studies of the material collected from the Nida River, which belongs to the left basin of the upper Vistula, and its left and right tributaries, the Maskalis and the Brzeźnica. These river channels are built of marl, limestone and gypsum deposits, postglacial sands, boulder clays of varying sandification as well as loess deposits of various depths, overlying limestone deposits. As is the case with the Bobrza and Gorzyczanka rivers, there are no natural sources of metal pollution. Farmland takes 69.3% of the catchment area, in which arable land accounts for 55.6%, orchards – for 1.3%, and meadows and pastures – for 12.4%. Forests, mostly coniferous and mixed, cover ca. 21.8% of the catchment area. The southern part of the catchment contains minor industrial centres (Jędrzejów, Pińczów, Sędziszów) and spa towns (Busko-Zdrój); gypsum and limestone are mined there as well, in the vicinity of Pińczów and Gacki [29, 30].

The Nida River and its tributaries receive municipal wastewater from three towns and industrial waste from plants representing industries such as mining, cement and lime, plaster goods, building materials, metallurgy and agro-food production, which to a large extent affects the waste quality. The agro-industrial nature of the catchment area causes farmland surface flows to affect surface water cleanliness. Particularly favourable conditions for fertiliser outwash are found in the central and southern parts of the catchment, given their intensive farming, impervious substratum, a dense river network and hilly terrain.

High metal amounts in the exchangeable form (20-40% of total sediment content) can indicate environmental acidification. In some authors' studies of bottom sediments [31-34], zinc content in the exchangeable form, for instance, was low and did not exceed 2-3% of this element's total content in the sediment. Its low mobility may have been caused by the pH of the analysed environments, which approximated 8.7, since at pH > 8 zinc is precipitated out of water [29].

The content of Fraction F2 of high bioavailability, or metals adsorbed on the extended surface of precipitated Fe and Mn oxides and hydroxides, amounted on average to 19% for Zn, 14% for Cd and 18% for Pb in studied sediments. However, the proportion of metal carbonate associations in bottom sediments was the lowest in the Nida River (8% Zn, 12% Cd and 11% Pb), and the highest in the Brzeźnica (26% Zn and Cd) and the Maskalis (17% Pb). Organic association content was the highest for lead (average of 40%), and the lowest for cadmium (16%), which may suggest for instance the occurrence of alkalisation processes in the environment, to which lead is subject as it turns into metalorganic compounds. Fraction F5, in turn, including analytes built into the crystal network of primary and secondary elements which are sediment components, indicates metals which are practically immobile, whose chemical compounds are passive, and biological ones – unavailable. The proportion of the residual fraction in the total zinc and lead content was small (2-11%), reaching 40% for cadmium in the sediment sampled in the Maskalis River [29].

Another example is the study of the Radomka River (a left tributary of the Vistula), and the Mleczna River (a tributary of the Radomka) [30]. The Radomka source substratum contains deposits of clayey resources and sandstone, iron ore (Jurassic siderites), and a phosphate belt. The predominant part of the Radomka valley holds shallow peat beds, and – occasionally – also brown coal. In its middle course, the river runs along a shallow marshy valley and then traverses the Kozienicka Primeval Woodland Complex. The Radomka catchment area is predominantly agricultural, over 50% being occupied by arable land, 22%– by woodland, 11% – of meadows and 6% – by orchards. Loose and loamy sands preponderate (43% of farmland area). Podsol soils account for 38.5%, sandy soils – for 30.5%, organogenic soils – for 11.2%, while brown soils, black earths and degraded alluvial soils take up ca. 6% each. The Radomka basin waters are polluted mainly by municipal wastewater, with the proportion of industrial wastewater not exceeding 10-20% [30].

The differences in zinc content and origin in sediment samples from the Radomka and Nida river basins have been reflected in different distributions of zinc among the determined chemical fractions. The basic characteristic of the zinc found in the Radomka basin sediment samples is the obvious relationship between the content of anthropogenic zinc and the proportion of the carbonate fraction. In a sample with natural zinc content, the carbonate fraction bonds 23.2% zinc, while in a sediment sample with heavy zinc pollution (267 mg·kg^{-1}), this fraction's proportion amounts to 88.4%. In samples collected at other locations, with middling zinc contents (24.8 mg·kg^{-1} and 158.7 mg·kg^{-1}), 45.8% and 83.0% of zinc is bonded with the carbonate fraction. The proportion of the less mobile oxide zinc fraction remains at stable levels in all studied samples, ranging from 3.0 to 10.6%. Determination results for the oxidisable fraction indicate different conditions for the formation of this fraction in the Radomka and Mleczna river channels. In the Radomka-Domaniów and Radomka-Bartodzieje samples, concentrations of this form of zinc are comparable: 3.53 mg·kg^{-1} and 2.06 mg·kg^{-1}, and it may be assumed that the zinc bonded to organic substance corresponds to this fraction. Higher contents of the oxidisable fraction, i.e. 6.7 mg·kg^{-1} and 9.2 mg·kg^{-1}, determined in the sediments of the heavily polluted Mleczna River waters, are rather the sum of the organic and sulphide forms. The formation of the latter is fostered by the oxygen deficit which occurs in the Mleczna River waters. Inert in the environment, the residual fraction bonds from 20.0% to 35.6% of zinc in the Radomka River sediments, and but a few per cent (3-5%) in the Mleczna River sediments. It must be stressed that in all the other samples (except for the sample with the highest zinc content) from both studied catchments, regardless of total zinc content, zinc content in the residual fraction remains at the level of a few mg·kg^{-1} [30].

The results of chemical fractionation of zinc in river sediments do not differ from the recently published results for other European river sediments. The study by Mossop and Davidson [35] gives the results of chemical fractionation of zinc in the White Cart River sediments (Great Britain) which contain 81 mg·kg^{-1} Zn: F(1)-Zn 42%, F(2)-Zn 13%, F(3)-Zn 19%, and the residual fraction – 26%. In the Ceruj River sediments (Romania) which contain from 120 mg·kg^{-1} to 650 mg·kg^{-1} of zinc, the carbonate fraction bonds from 30% to 70% of zinc. In sediments with the highest zinc content, Fraction F(1) predominates [36]. Similar results have been obtained by Helios-Rybicka et al. [37] for the heavily polluted Odra River sediments, the ion-exchangeable and carbonate fractions in the sediment grain fraction <0.63 μm bonding up to 50% of zinc. In the Pisuerga River sediments (Spain), which receives industrial and municipal wastewater, depending on the sampling site, the main zinc fractions are the mobile (34%), the reducible (40%) or the oxidisable (53%) [38].

Similarly, bottom sediment studies in Asian rivers suggest considerable impact of human use of rivers and catchment areas on physical and chemical properties of the sediments, including the occurrence of heavy metal forms. An example is the fractionation of bottom sediments from the Pearl River, the largest river system flowing into the South China Sea, and the analysis of the obtained extracts for Pb, Cu and Zn [39]. The Pearl River delta is a very large agglomeration located in the south of China, featuring large numbers of factories, inexpensive manufacturing plants, multi-million cities and environmental pollution. The sediment samples were taken at various depths, from seventeen sites, fractionated with Tessier's method and subsequently analysed for Pb, Cu and Zn content. The proportions of Pb associated with various fractions were as follows: residual > Fe-Mn oxide > organic > carbonate > exchangeable. Yet fraction F5 (residual) is of little significance in bottom sediments on the west of the river's delta, and the highest amounts of Pb have been recorded in fractions F3 and F4. The varied lead distribution in various samples is conditioned by human activity and is of anthropogenic character. Zinc distribution on the surface and in particular layers taken from various depths varies. The majority of Zn is found in Fraction V. Smaller amounts have been noted, respectively, in fractions F3 (Fe-Mn oxide) and F4 (organic), while the lowest comparable amounts have been found in exchangeable and carbonate fractions. A similar distribution has been found for Cu, whose content was as follows: residual > organic > Fe-Mn oxide > carbonate > exchangeable. High Cu content in fractions F4 and F5 causes this element to be less mobile than Pb and Zn [39].

Another fraction distribution in bottom sediments has been determined through fractionation of bottom sediment samples of the Narmada River in India [40]. The material has been collected at twelve sites along the entire river length and analysed for Fe, Mn, Ni, Zn, Cr, Cu, Pb and Cd content. The Narmada river channel is built primarily of basalt type rock with a red soil cover. The river is fed by several other rivers; it flows across urbanised areas, and is traversed by communications routes; additionally, it has several dams and dykes – accordingly, the areas located within its range are classified as degraded.

The copper fractionation profile indicates that a major portion is bonded to the residual fraction and the complexation with organic matter fraction. A substantial proportion of copper is also found in reducible (Fe–Mn oxide) and organic fractions, probably due to its more pronounced tendency for complexation with organic matter. The nickel fractionation profile indicates that in general nickel is associated in decreasing amounts in the following fraction order: residual > exchangeable > reducible > bonded to carbonate > oxidisable fraction. The chromium fractionation profile indicates that more than 50% of chromium is associated with first three fractions (exchangeable, carbonate bound and reducible) at most of the sites and can enter the food chain. The lead fractionation profile indicates that more than 90% of lead is associated with the residual fraction. A small portion (1-4%) is also associated with the exchangeable fraction, which indicates an anthropogenic source from municipal and industrial discharges. The cadmium fractionation profile indicates that a major portion of cadmium is associated with the residual fraction. About 25% of cadmium is also associated with the first three fractions (exchangeable, carbonate and reducible) and may be easily remobilised by changes in environmental conditions. The association of cadmium with the exchangeable fraction indicates the dominance of anthropogenic sources through atmospheric deposition and municipal discharges. The toxic nature of cadmium and its association with these fractions may cause deleterious effects to aquatic life [40].

Heavy metal binding forms for Cu, Zn and Pb have been determined at four representative sediment sampling sites in the canals of Delft (The Netherlands), with the use of Tessier's chemical extraction method [41]. The different heavy metal binding fractions in the sediment indicate that Cu is mostly (51–83%) present in the F4 (organic/sulphidic) phase, followed by the residual phase (F5), which was relatively prominent in the "background station" (highly polluted). Contributions of the remaining, labile fractions F1 - F3 were low, i.e. max. 5% in total. The dominant binding form for Zn was found to be Fraction F3 ("bound to Fe/Mn oxides"; ca. 60%), with most of the remainder in fractions F4 and F5. Finally, for Pb, the overall picture was more varied, with dominant binding forms being either fraction F5 (unpolluted and highly polluted stations) or F3 (slightly and medium polluted stations), and relatively high proportions of other binding forms. Thus, the average proportion of Fraction F1 (exchangeable) amounted to 15%, compared with 2.4% for Cu and Zn [41].

In contrast to others [38, 42], Kelderman and Osman [41] did not find significant ($p > 0.05$) correlations between loss on ignitron (LOI) and cation exchange capacity (CEC) on the one hand, and heavy metal contents on the other. The same holds true for the relationship between $CaCO_3$ and heavy metal content, which is in agreement with other researchers' findings [38, 42, 43]. Mixed observations have been reported for correlations between Fe and Mn and heavy metal content [43, 44]. However, the relatively high R^2 values (i.e. between 0.65 and 0.9) are an indication of marked importance of Fe and Mn phases for heavy metal binding (mainly as (hydr)oxides) [41].

Interesting results were obtained by studying the sediments of Bobrza River. The samples were taken from two different places characterized by different degree of acidity and alkalinity. The 1st point located in the impact zone of the Kielce Pump Factory and Formaster Company. The soils in the area are acidic and count among podsol soils, formed primarily out of glacier accumulation formations or on non-carbonate Triassic sandstones [50, 51].

In the 2nd point a chemical composition of cement and lime dust is rather constant and mainly depends on the composition of raw materials used in technological process, production technology, way of clinker burning and type of additives enriching the cement. The dust emitted from calcareous institutions cement plant contains mainly CaO, SiO_2, Al_2O_3, K_2O and has high values of pH_{KCl}. The exhaust dust contains also heavy metals including Zn (173.0 mg·kg^{-1}), Pb (140.0 mg·kg^{-1}) and Cd (3.0 mg·kg^{-1}) which causes enrichment of the soil and water environment with compounds of these elements [50, 51]. The presence of high metal contents in bottom sediments in the form of carbonate and oxide fractions is the result of environmental alkalinisation and deposition of dusts containing alkaline metal oxides together with non-ferrous metal oxides.

Different types of land development at the level of a single river (Fig. 1-3) make it impossible to develop one general scheme for metal migration in the water - bottom sediment system. The number of factors, particularly anthropogenic ones, which determine the ion composition of water, the amount and type of the suspension, oxygenation extent, presence of microorganisms, and - consequently - forms in which metals are found in the ecosystem is too high to enable a definite statement that under certain conditions remobilisation or precipitation of a poorly soluble compound of a given metal will occur.

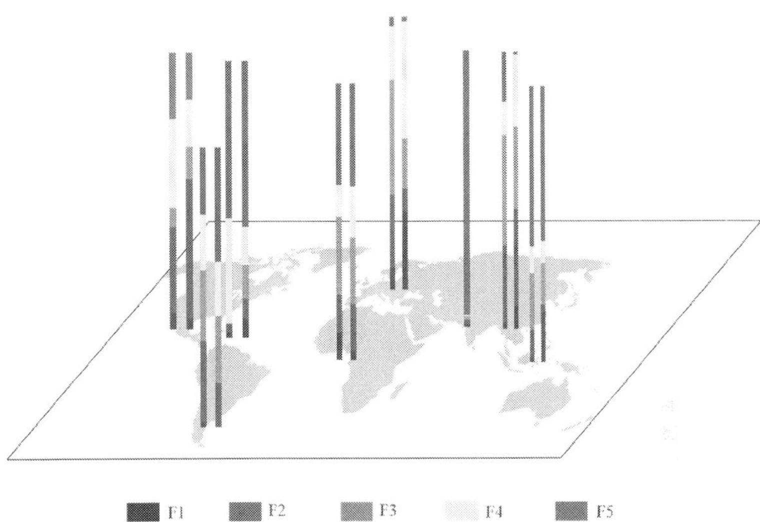

Fig. 1. The lead forms in bottom sediments in selected rivers of the world [on the basis of: 27, 29, 40, 45-49]

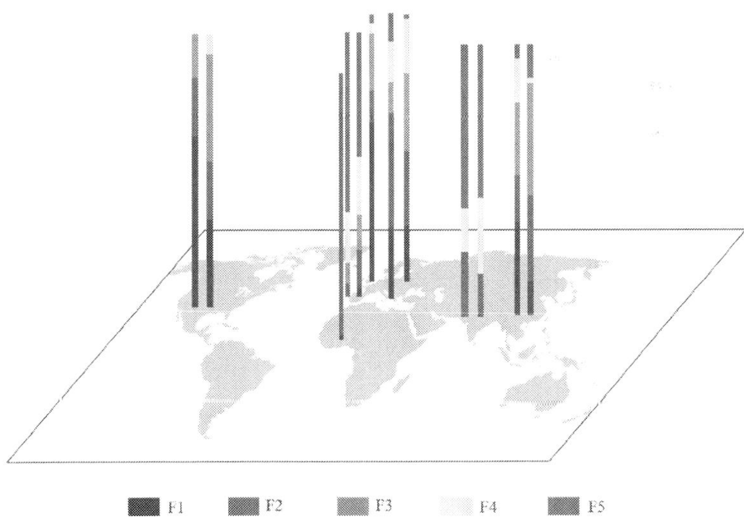

Fig. 2. The chromium forms in bottom sediments in selected rivers of the world [on the basis of: 51-55, 57-60].

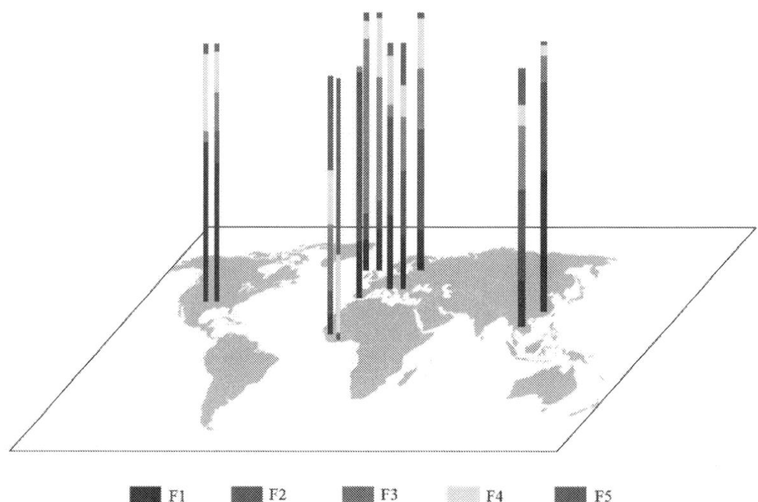

Fig. 3. The cooper forms in bottom sediments in selected rivers of the world [on the basis of: 51-59].

3. Conclusions

Surface waters, depending on their geological structure, substratum type and land development, are characterised by varied chemical composition. Metals such as Cd, Cr, Cu, Fe, Pb, Ni or Zn pass into water as a result of natural processes, as well as direct and indirect human activity. The number of sources from which substances originate, chemical properties of individual entities, their capacity for adsorption on solid particles, and forming hydrated ions or ion pairs, together with biological, physical and chemical conditions of the aquatic environment, all affect the variety of metal forms found in surface waters. Elements, while migrating from the environment, are deposited in bottom sediments, where they are temporarily immobilised and may pose a hazard to biological life in a given ecosystem.

The knowledge of the quantity of metals in waters and bottom sediments may be used for ultimate chemical assessment of an aquatic environment and serve as an indicator of the geochemical status of a given catchment as well as the spread of pollution. Due to their properties, bottom sediments which are the main component of the elemental cycle, and the centre of accumulation, chemical transitions, and decomposition of toxic compounds which enter the aquatic environment, are a rather representative indicator of long-term changes in aquatic ecosystem pollution. They may be used as indicators of biological and chemical change in the aquatic environment.

However, they may also pose a hazard of secondary pollution to the ecosystem as a result of desorption and passage of substances accumulated on the surface of solids into water. The soluble complex compounds which are then formed with organic and inorganic ligands present in rivers may cause secondary release of heavy metals in various forms, more or less dangerous to living organisms.

Consequently, each aquatic ecosystem ought to receive individual treatment, allowances being made for both natural factors (including geochemical background, or climate) and anthropogenic factors (land development in the catchment and the river itself). In order to develop schemes for metal migration or the probability of secondary pollution of the ecosystem which can enhance water resources management, it is necessary to conduct research in representative river segments, comparable in terms of natural factors but different in terms of anthropogenic ones.

4. Acknowledgements

The author would like to thank the State Committee for Scientific Research (KBN, Poland) Grants No. N N305 306635 and   The Jan Kochanowski University of Humanities and Sciences in Kielce for providing funding for the project.

5. References

[1] J. Namieśnik and A. Rabajczyk, *Chem Speciation Bioavailability*, 2010, 22(1), 1.
[2] D.R. Turner, in *Metal Speciation and Bioavailability in Aquatic Systems*, ed. A. Tessier and D.R. Turner, IUPAC, Wiley, Chichester, 1995, pp. 149-203.
[3] J. Kalembkiewicz and E. Sočo, *Pol J Environ Stud*, 2002, 11(3), 245.
[4] Rabajczyk, *Ecol Chem Eng S1*, 2006, 13, 167.
[5] A.J. Breckel, S. Emerson and L.S. Balistrieri, *Continental Shelf Research*, 2005, 25, 1321–1337.
[6] Morford, J.L. and Emerson, S., *Geochimica et Cosmochimica Acta*, 1999, 63, 1735–1750.
[7] J. Thompson, M. Carpenter, S. Colley, T.R.S. Wilson, H. Elderfield and H. Kennedy, *Geochimica et Cosmochimica Acta*, 1984, 48(10), 1935-1948.
[8] S. Gatehouse, D.W. Russell and J.C. Van Moort, *J Geochem Explor*, 1977, 8, 483.
[9] Tessier, P.G.C. Campbell and M. Bisson, *Anal Chem*, 1979, 51, 844.
[10] G. Sposito, L.J. Lund and A.C. Chang, *Soil Sci Soc Am J*, 1982, 46, 260.
[11]A.M. Ure, P. Quevauviller, H. Muntau and B. Griepin, *Int J Environ Anal Chem*, 1993, 51, 135.
[12] G.E.M. Hall, G. Gauthier, J.C. Pelchat, P. Pelchat and J.E. Vaive, *J Anal At Spectrom*, 1996, 11, 787.
[13] R.M. Cornell and U. Schwertmann, *Cosmochimica Acta*, 1996, 48, 1935–1948.
[14] J.J. Morgan, in: A. Sigel, H. Sigel (eds). *Metal Ions in Biological Systems*, New York: Marcel Dekker, 2000, pp 1-34.
[15] T.M. Scot, in: *Environmental Catalysis*, V.H. Grassian (ed.) CRC Press: Boca Raton, 2005, Chap. 4, pp 61-78.
[16] N.L., Dollar C.J. Souch, G.M. Filippelli and M. Mastalerz, *Environ Sci Technol*, 2001, 35, 3608-3615.
[17] B.M. Petrunic, K.T.B. MacQuarrie and T.A. Al, *Journal of Hydrology*, 2005, 301, 163-181.
[18] R.Y. Li, H. Yaxg, Z.G. Zhou, J.J. Lü, X.H. Shao and F. Jin, *Pedosphere*, 2007, 17(2), 265.
[19] M. Sager, R. Pucsko and R. Belocky, *Arch Hydrobiol Suppl*, 1990, 84, 37.
[20] M. Kļaviņš, A. Briede, V. Rodinov, I. Kokorīte, E. Parele and I. Kļaviņa, *Sci Total Environ*, 2000, 262, 175.
[21] S. Lee, J.W. Moon and H.S. Moon, *Environ Geochem Health*, 2003, 25, 433.
[22] C.K. Jain, *Wat Res*, 2004, 38, 569.
[23] M.R. Lasheen and N.S. Ammar, *Environmentalist*, 2009, 29, 8.
[24] Water Policy Program. Survey of Nile System Pollution Sources Report No. 64. APRP-Water Policy Activity, 2002.
[25] K.P. Singh, D. Mohan, V.K. Singh and A. Malik, *J Hydrol*, 2005, 312, 14.
[26] H. Akcay, A. Oguz and C. Karapire, *Wat Res*, 2003, 37, 813.
[27] O.A.A. Eletta and F.A. Adekola, *Ife Journal of Science*, 2005, 7(1), 139.

[28] Rabajczyk, *Pol J of Environ Stud*, 2009, 178(2B), 35.

[29] J. Namieśnik and A. Rabajczyk, *Pol J. of Environ. Stud.*, 2010 (in press).

[30] R. Świetlik, A. Rabajczyk and M. Trojanowska, *Geological Review*, 2009, 57(12), 1101.

[31] K. Loska, D. Wiechuła and J. Cebula, *Pol J Environ Stud*, 2000, 9, 523.

[32] J.J. Vicente-Martorell, M.D. Galindo-Riaño, M. García-Vargas and M.D. Granado-Castro, *J Hazard Mater*, 2009, 162, 823.

[33] D. Wiechuła, J. Kwapulinski and A. Paukszto, *Proc. Int. Symp. Heavy Metals in Environment*, Toronto, 1993, 1, 193.

[34] J. *Zerbe*, T. Sobczynski, H. Elbananowska and J. Siepak, *Pol J Environ Stud*, *1999*, 8(5), 331.

[35] K.F. Mossop and C.M. Davidson, *Anal Chim Acta*, 2003, 478, 111.

[36] G.D. Vasile, M. Nicolau and L. Vladescu, *Environ Monit Assess*, 2010, 160(1-4), 71.

[37] E. Helios-Rybicka, E. Adamiec and U. Aleksander-Kwaterczak, Limnologica(Jena), 2005, 35, 185.

[38] R. Pardo, E. Barrado, L. Perez and M. Vega, *Wat Res*, 1990, 24, 373.

[39] X. Li, Z. Shen, O.W.H. Wai and Y.S. Li, *Marine Pollution Bulletin*, 2001, 42(30), 215.

[40] C.K. Jain, H. Gupta and G.J. Chakrapani, *Environ Monit Assess*, 2008, 141, 35.

[41] P. Kelderman and A.A. Osman, *Wat Res*, 2007, 41, 4251.

[42] F. Howari and K.M. Banat, *Water Air Soil Pollut*, 2001, 132(1-2), 43.

[43] B. Šurija and M. Branica, *Sci Tot Environ*, 1995, 170, 101.

[44] U. Fôrstner and G.T.W. Wittman, Metal Pollution in the Aquatic Environment (2nd edn). Springer Verlag, Berlin, 1983.

[45] C.S.C Wong, S.C. Wu, N.S. Duzgoren-Aydin, A. Aydin and M.H. Wong, *Environ Poll*, 2007, 145(2), 434.

[46] R. Segura, V. Arancibia, M.C. Zúñiga, P. Pastén, *J Geochem Explor*, 2006, 91, 71.

[47] S. Olivares-Rieumont, D. de la Rosa, L. Lima, D.W. Graham, K.D. Alessandro, J. Borroto, F. Martínez and J. Sánchez, *Wat Res*, 2005, 39, 3945.

[48] F. Arcega-Cabrera, M.A. Armienta, L.W. Daesslé, S.E. Castillo-Blum, O. Talavera and A. Dótor, *Appl Geochem*, 2009, 24, 162.

[49] N.D. Takarina, D.R. Browne and M.J. Risk, *Mar Poll Bull*, 2004, 49, 854.

[50] A.Rabajczyk, *Polish J. of Environ. Stud.*, 2009, 178(2B), 35-41.

[51] A.Rabajczyk, Central European Journal of Chemistry, 2011, 9(2), 326.

[52] M.J.B. Segarra, R. Prego, M.J. Wilson, J. Bacon and J. Santos-Echeandia, *Scientia Marina*, 72(1), 2008, 119-126.

[53] B. M. Wufem, A. Q. Ibrahim, N. S. Gin, M.A. Mohammed, E.O. Ekanem and M.A. Shibdawa, *Global Journal Of Environmental Sciences*, 2009, 8(2), 55-63.

[54] P.C. Ryan, A.J. Wall, S. Hillier and L. Clark, *Chemical Geology*, 2002, 184, 337-357.

[55] J. Carter, D.E. Walling, P.N. Owens and G.J.L. Leeks, *UK. Hydrol Processes*, 2006, 20, 3007–3027.

[56] S. Lee, J.W. Moon and H.S. Moon, *Environmental geochemistry and health*, 2003, 25, 433-452.

[57] K. Fytianos and A. Lourantou, *Environment International*, 2004, 30, 11-17.

[58] N.D. Takarina, D.R. Browne and M.J. Risk, *Marine Pollution Bulletin*, 2004, 49, 854-874.

[59] Q.S. Li, Z.F. Wu, B. Chu, N. Zhang, S.S. Cai and J.H. Fang, *Environmental Pollution*, 2007, 149, 158-164.

[60] K.P. Singh, D. Mohan, V.K. Singh and A. Malik, *Journal of Hydrology*, 2005, 312, 14-27.